AF655413

Fundamentals of
COORDINATE GEOMETRY

Fundamentals of COORDINATE GEOMETRY

Pradeep Aggarwal
J.K. Watson

KNOWLEDGE BAKERS

KNOWLEDGE BAKERS

Cataloging in Publication Data--DK
Courtesy: D.K. Agencies (P) Ltd. <docinfo@dkagencies.com>

Aggarwal, Pradeep, author.
Fundamentals of coordinate geometry / Pradeep Aggarwal, J.K. Watson.
pages cm
Includes bibliographical references and index.
ISBN 9789391502201

1. Coordinates. 2. Geometry, Analytic. I. Watson, J. K. (Writer on coordinate geometry), author. II. Title. III. Title: Coordinate geometry.

LCC QA556.A34 2023 | DDC 516.16 23

© 2023 AUTHORS

ISBN: 978-93-91502-20-1

9 789391 502201

All rights reserved under International Copyright Conventions. No part of the book or part thereof, including title of the book, be reproduced in any printed form or language without written permission of the author and the publishers.

Published by : **Knowledge Bakers**
Pune, India
Email: knowledgebakers2019@gmail.com

Digitally Printed at : **Replika Press Pvt. Ltd.**

Preface

The most common coordinate system to use is the Cartesian coordinate system, where each point has an x-coordinate representing its horizontal position, and a y-coordinate representing its vertical position. These are typically written as an ordered pair (x, y). This system can also be used for three-dimensional geometry, where every point in Euclidean space is represented by an ordered triple of coordinates (x, y, z). *Other* coordinate systems are possible. On the plane the most common alternative is polar coordinates, where every point is represented by itsradius r from the origin and its angle è. In three dimensions, common alternative coordinate systems include cylindrical coordinates and spherical coordinates. The reduction of geometry to algebra requires the notion of a transformation group. The transformation group supplies two essential ingredients. First it is used to define the notion of equivalence in the geometry in question. For example, in Euclidean geometry, two triangles are congruent if there is distance preserving transformation carrying one to the other and they are similar if there is a similarity transformation carrying one to the other. Secondly, in each kind of geometry there are normal form theorems which can be used to simplify coordinate proofs. In analytic geometry, the plane is given a coordinate system, by which every point has a pair of real number coordinates.

Transformations are applied to parent functions to turn it into a new function with similar characteristics. For example, the parent function y=1/x has a horizontal and a vertical asymptote, and occupies the first and third quadrant, and all of its transformed forms have one horizontal and vertical asymptote, and occupies either the 1st and 3rd or 2nd and 4th quadrant. In general, if $y = f(x)$, then it can be transformed into $y = af(b(x - k)) + h$. In the new transformed function, a is the factor that vertically stretches the function if it is greater than 1 or vertically compresses the function if it is less than 1, and for negative a values, the function is reflected in the x-axis. The b value compresses the graph of the function horizontally if greater than 1 and stretches the function horizontally if less than 1, and like a, reflects the function in the y-axis when it is negative. The k and h values introduce translations, h, vertical, and k horizontal. Positive h and k values mean the function is translated to the positive end of its axis and negative meaning translation towards the negative end. One type of intersection which is widely studied is the intersection of a geometric object with the x and y coordinate axes. The intersection of a geometric object and the y-axis is called the y-intercept of the object. The intersection of a geometric object and the x-axis is called the x-intercept of the object.For the line y=mx+b, the parameter b specifies the point where the line crosses the y axis. Depending on the context, either b or the point (0,b) is called the y-intercept.

The present book provides an authoritative review of many aspects of this subject. This book will be useful to teachers and students.

Dr Pradeep Aggarwal
Dr J.K. Watson

CONTENTS

Preface v

1. **Coordinate System** 1

Number Line – Cartesian Coordinate System – Definitions – Octant (Solid Geometry) – Quadrant (Plane Geometry) – Cartesian Formulas for the Plane – Alternate Viewpoints – Orientation on Manifolds – Axes Conventions – Frames Mounted on Vehicles – Right-hand Rule

2. **Polar Coordinate System** 29

History – Uniqueness of Polar Coordinates – Spherical Coordinate System – Coordinate System Conversions – Homogeneous Coordinates – Duality (Projective Geometry) – Duality as a Mapping – Plücker Coordinates– Manifold – Motivational Examples – Charts, Atlases, and Transition Maps – Manifolds with additional structure

3. **6-Sphere Coordinates** 85

Abscissa – Affine Coordinate System – Alpha-numeric Grid – Astronomical Coordinate Systems – Bipolar Coordinates – Bipolar Cylindrical Coordinates – Bispherical Coordinates – Hyperbolic Coordinates

4. **Line Coordinates** 96

Lines in the Plane – Log-Polar Coordinates – Some Important Equations in Log-polar Coordinates – Orthogonal Coordinates – Basis Vectors – Vector Calculus – Parabolic Cylindrical Coordinates – Two-centre Bipolar Coordinates – Darboux's Theorem – Isothermal Coordinates – Normal Coordinates

5. **Derivations of the Quadratic Formula** 117

By Completing the Square – Other Methods of Root Calculation – Trigonometric Functions – Conic Section – History – Matrix Representation of Conic Sections – Plane (Geometry) – Euclidean Geometry – Notation and Terminology – Later Work

6. Treatment of Infinity **152**

Infinite Objects – Logical Basis – Planes Embedded in 3-dimensional Euclidean Space – Line (Geometry) – Rotation (Mathematics) – Circle Group – Topological and Analytic Structure – Isomorphisms – Rotation Group SO(3)

7. The Hypersphere of Rotations **167**

Visualising the Hypersphere – Rotation Formalisms in Three Dimensions – Axis-angle Representation – Rotation Vector – Euler Angles – Quaternions and Spatial Rotation – Explaining Quaternions' Properties with Rotations – Quaternions and Other Representations of Rotations – Conversion Between Quaternions and Euler Angles

8. Rotation Matrix **195**

In Two Dimensions – Ambiguities – Decompositions – Lie Theory – Infinitesimal Rotations – Conversions

9. Uniform Random Rotation Matrices **216**

Aircraft Principal Axes – Relationship with Other Systems of Axes – Proper Euler Angles – Properties – Applications – Cylinder (Geometry) – Cylindric Section – Other Types of Cylinders

10. Surface Area of a Sphere **227**

Equations in R^3 – Description – Generating Random Points – Specific Spheres – Generalisation to Metric Spaces – Topology – Spherical Geometry – Relation to Euclid's Postulates – Eleven Properties of the Sphere – Degree (Angle) – Subdivisions - Alternative Units - Angle – Measuring Angles - Identifying Angles - Types of Angles – Angles Between Curves

11. Dot Product and Generalization **252**

Angles Between Subspaces – Cone (Geometry) – Other Mathematical Meanings – Conical Surface – Surface of Constant Width – Projective Geometry – Axioms of Projective Geometry – Hilbert's Third Problem – Dandelin Spheres

Bibliography **269**

Index **271**

Chapter

1

Coordinate System

In geometry, a coordinate system is a system which uses one or more numbers, or coordinates, to uniquely determine the position of a point or other geometric element on a manifold such as Euclidean space. The order of the coordinates is significant and they are sometimes identified by their position in an ordered tuple and sometimes by a letter, as in 'the x-coordinate'. In elementary mathematics the coordinates are taken to be real numbers, but may be complex numbers or elements of a more abstract system such as a commutative ring. The use of a coordinate system allows problems in geometry to be translated into problems about numbers and *vice versa*; this is the basis of analytic geometry.

An example in everyday use is the system of assigning longitude and latitude to geographical locations. In physics, a coordinate system used to describe points in space is called a frame of reference.

Number Line

In basic mathematics, a number line is a picture of a straight line on which every point is assumed to correspond to a real number and every real number to a point. Often the integers are shown as specially-marked points evenly spaced on the line. Although this image only shows the integers from –9 to 9, the line includes all real numbers, continuing forever in each direction, and also numbers not marked that are between the integers. It is often used as an aid in teaching simple addition and subtraction, especially involving negative numbers.

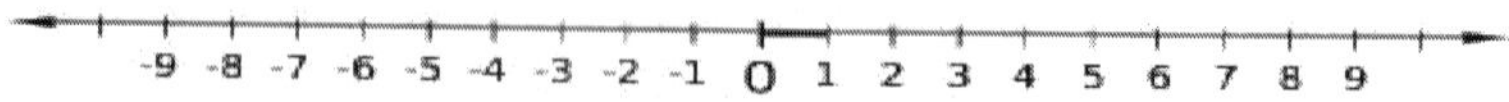

It is divided into two symmetric halves by the origin, i.e. the number zero. In advanced mathematics, the expressions real number line, or real line are typically used to indicate the above-mentioned concept that every point on a straight line corresponds to a single real number, and vice versa.

Drawing the Number Line

The number line is usually represented as being horizontal. Customarily, positive numbers lie on the right side of zero, and negative numbers lie on the left side of zero. An arrowhead on either end of the drawing is meant to suggest that the line continues indefinitely in the positive and negative real numbers, denoted by □ . The real numbers consist of irrational numbers and rational numbers, as well as the integers, whole numbers, and the natural numbers (the counting numbers).

A line drawn through the origin at right angles to the real number line can be used to represent the imaginary numbers. This line, called imaginary line, extends the number line to a complex number plane, with points representing complex numbers.

Cartesian Coordinate System

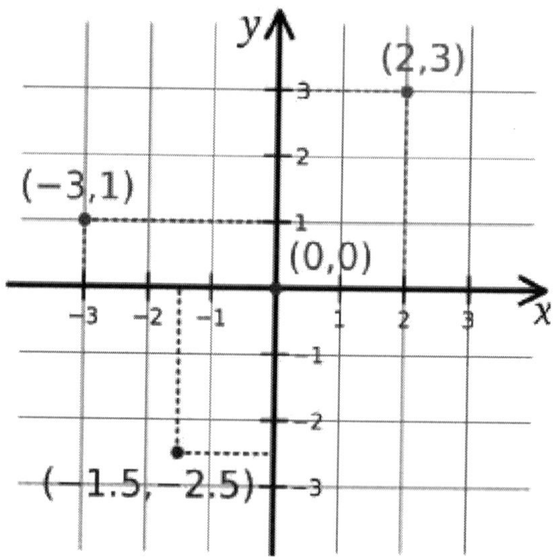

Figure: *Illustration of a Cartesian coordinate plane. Four points are marked and labeled with their coordinates: (2, 3) in green, (–3, 1) in red, (–1.5, –2.5) in blue, and the origin (0, 0) in purple.*

A Cartesian coordinate system specifies each point uniquely in a plane by a pair of numerical coordinates, which are the signed distances from the point to two fixed perpendicular directed lines, measured in the same unit of length. Each reference line is called a *coordinate axis* or just *axis* of the system, and the point where they meet is its *origin*, usually at ordered pair (0,0). The coordinates can also be defined as the positions of the perpendicular projections of the point onto the two axes, expressed as signed distances from the origin.

One can use the same principle to specify the position of any point in three-dimensional space by three Cartesian coordinates, its signed distances to three mutually perpendicular planes (or, equivalently, by its perpendicular projection onto three mutually perpendicular lines). In general, one can specify a point in a space of any dimension *n* by use of *n* Cartesian coordinates, the signed distances from *n* mutually perpendicular hyperplanes.

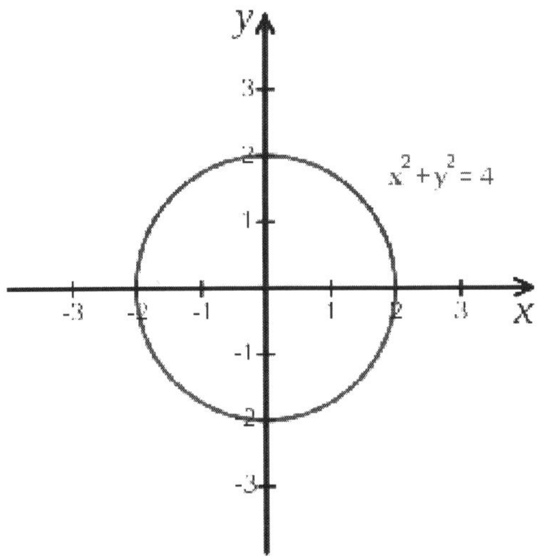

Figure: *Cartesian coordinate system with a circle of radius 2 centreed at the origin marked in red. The equation of a circle is* $(x - a)^2 + (y - b)^2 = r^2$ *where a and b are the coordinates of the centre (a, b) and r is the radius.*

The invention of Cartesian coordinates in the 17th century by René Descartes (Latinized name: *Cartesius*) revolutionised mathematics by providing the first systematic link between Euclidean geometry and algebra. Using the Cartesian coordinate system, geometric shapes (such as curves) can be described by Cartesian equations: algebraic equations involving the coordinates of the points lying on the shape. For example, a circle of radius 2 may be described as the set of all points whose coordinates x and y satisfy the equation $x^2 + y^2 = 4$.

Cartesian coordinates are the foundation of analytic geometry, and provide enlightening geometric interpretations for many other branches of mathematics, such as linear algebra, complex analysis, differential geometry, multivariate calculus, group theory, and more.

A familiar example is the concept of the graph of a function. Cartesian coordinates are also essential tools for most applied disciplines that deal with geometry, including astronomy, physics, engineering, and many more. They are the most common coordinate system used in computer graphics, computer-aided geometric design, and other geometry-related data processing.

History

The adjective *Cartesian* refers to the French mathematician and philosopher René Descartes (who used the name *Cartesius* in Latin).

The idea of this system was developed in 1637 in writings by Descartes and independently by Pierre de Fermat, although Fermat also worked in three dimensions, and did not publish the discovery.

Both authors used a single axis in their treatments and have a variable length measured in reference to this axis. The concept of using a pair of axes was introduced in later work by commentators who were trying to clarify the ideas contained in Descartes' *La Géométrie*.

The development of the Cartesian coordinate system would play an intrinsic role in the development of the calculus by Isaac Newton and Gottfried Wilhelm Leibniz.

Nicole Oresme, a French cleric and friend of the dauphin (later to become King Charles V) of the 14th Century, used constructions similar to Cartesian coordinates well before the time of Descartes and Fermat.

Many other coordinate systems have been developed since Descartes, such as the polar coordinates for the plane, and the spherical and cylindrical coordinates for three-dimensional space.

Definitions

Number Line

Choosing a Cartesian coordinate system for a one-dimensional space—that is, for a straight line—means choosing a point O of the line (the origin), a unit of length, and an orientation for the line. An orientation chooses which of the two half-lines determined by O is the positive, and which is negative; we then say that the line "is oriented" (or "points") from the negative half towards the positive half. Then each point p of the line can be specified by its distance from O, taken with a + or – sign depending on which half-line contains p.

A line with a chosen Cartesian system is called a number line. Every real number, whether integer, rational, or irrational, has a unique location on the line. Conversely, every point on the line can be interpreted as a number in an ordered continuum which includes the real numbers.

Cartesian Coordinates in Two Dimensions

The modern Cartesian coordinate system in two dimensions (also called a rectangular coordinate system) is defined by an ordered pair of perpendicular lines (axes), a single unit of length for both axes, and an orientation for each axis.

(Early systems allowed "oblique" axes, that is, axes that did not meet at right angles.) The lines are commonly referred to as the x and y-axes where the x-axis is taken to be horizontal and the y-axis is taken to be vertical.

The point where the axes meet is taken as the origin for both, thus turning each axis into a number line. For a given point P, a line is drawn through P perpendicular to the x-axis to meet it at X and second line is drawn through P perpendicular to the y-axis to meet it at Y.

The coordinates of P are then X and Y interpreted as numbers x and y on the corresponding number lines. The coordinates are written as an ordered pair (x, y).

The point where the axes meet is the common origin of the two number lines and is simply called the *origin*. It is often labeled O and if so then the axes are called Ox and Oy.

A plane with x and y-axes defined is often referred to as the Cartesian plane or xy plane. The value of x is called the x-coordinate or abscissa and the value of y is called the y-coordinate or ordinate.

The choices of letters come from the original convention, which is to use the latter part of the alphabet to indicate unknown values. The first part of the alphabet was used to designate known values.

In the Cartesian plane, reference is sometimes made to a unit circle or a unit hyperbola.

Cartesian Coordinates in Three Dimensions

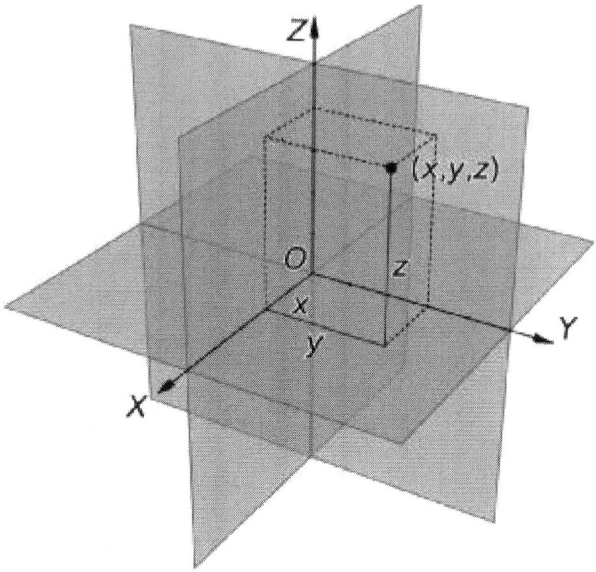

***Figure:** A three dimensional Cartesian coordinate system, with origin O and axis lines X, Y and Z, oriented as shown by the arrows. The tick marks on the axes are one length unit apart. The black dot shows the point with coordinates X = 2, Y = 3, and Z = 4, or (2,3,4).*

Choosing a Cartesian coordinate system for a three-dimensional space means choosing an ordered triplet of lines (axes), any two of them being perpendicular; a single unit of length for all three axes; and an orientation for each axis.

As in the two-dimensional case, each axis becomes a number line. The coordinates of a point p are obtained by drawing a line through p perpendicular to each coordinate axis, and reading the points where these lines meet the axes as three numbers of these number lines.

Alternatively, the coordinates of a point p can also be taken as the (signed) distances from p to the three planes defined by the three axes.

If the axes are named x, y, and z, then the x coordinate is the distance from the plane defined by the y and z axes. The distance is to be taken with the + or – sign, depending on which of the two half-spaces separated by that plane contains p. The y and z coordinates can be obtained in the same way from the (x, z) and (x, y) planes, respectively.

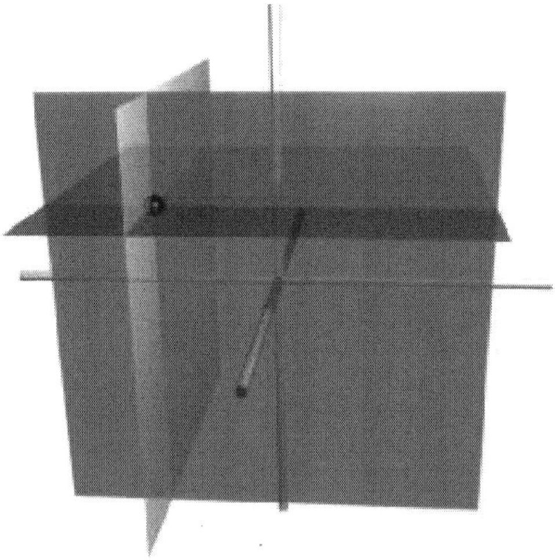

Figure: *The coordinate surfaces of the Cartesian coordinates (x, y, z). The z-axis is vertical and the* x-*axis is highlighted in green. Thus, the red plane shows the points with* $x = 1$*, the blue plane shows the points with* $z = 1$*, and the yellow plane shows the points with* $y = -1$*. The three surfaces intersect at the point P (shown as a black sphere) with the Cartesian coordinates (1, –1, 1).*

Generalisations

One can generalise the concept of Cartesian coordinates to allow axes that are not perpendicular to each other, and/or different units along each axis. In that case, each coordinate is obtained by projecting the point onto one axis along a direction that is parallel to the other axis (or, in general, to the hyperplane defined by all the other axes). In those oblique coordinate systems the computations of distances and angles is more complicated than in standard Cartesian systems, and many standard formulas (such as the Pythagorean formula for the distance) do not hold.

Notations and Conventions

The Cartesian coordinates of a point are usually written in parentheses and separated by commas, as in (10,5) or (3,5,7). The origin is often labelled with the capital letter *O*. In analytic geometry, unknown or generic coordinates are often denoted by the letters *x* and *y* on the plane, and *x*, *y*, and *z* in three-dimensional space. *w* is often used for four-dimensional space, but the rarity of such usage precludes concrete convention here. This custom comes from an old convention of algebra, to use letters near the end of the alphabet for unknown values (such as were the coordinates of points in many geometric problems), and letters near the beginning for given quantities. These conventional names are often used in other domains, such as physics and engineering. However, other letters may be used too. For example, in a graph showing how a pressure varies with time, the graph coordinates may be denoted t and P. Each axis is usually named after the coordinate which is measured along it; so one says the x-axis, the y-axis, the t-axis, etc.

Another common convention for coordinate naming is to use subscripts, as in x_1, x_2, ... x_n for the n coordinates in an n-dimensional space; especially when n is greater than 3, or variable.

Some authors (and many programmers) prefer the numbering x_0, x_1, ... x_{n-1}. These notations are especially advantageous in computer programming: by storing the coordinates of a point as an array, instead of a record, one can use iterative commands or procedure parametres instead of repeating the same commands for each coordinate. In mathematical illustrations of two-dimensional Cartesian systems, the first coordinate (traditionally called the abscissa) is measured along a horizontal axis, oriented from left to right.

The second coordinate (the ordinate) is then measured along a vertical axis, usually oriented from bottom to top.

However, in computer graphics and image processing one often uses a coordinate system with the y axis pointing down (as displayed on the computer's screen). This convention developed in the 1960s (or earlier) from the way that images were originally stored in display buffers. For three-dimensional systems, the z axis is often shown vertical and pointing up (positive up), so that the x and y axes lie on a horizontal plane. If a diagram (3D projection or 2D perspective drawing) shows the x and y axis horizontally and vertically, respectively, then the z axis should be shown pointing "out of the page" towards the viewer or camera.

In such a 2D diagram of a 3D coordinate system, the z axis would appear as a line or ray pointing down and to the left or down and to the right, depending on the presumed viewer or camera perspective.

In any diagram or display, the orientation of the three axes, as a whole, is arbitrary. However, the orientation of the axes relative to each other should always comply with the right-hand rule, unless specifically stated otherwise. All laws of physics and math assume this right-handedness, which ensures consistency. For 3D diagrams, the names "abscissa" and "ordinate" are rarely used for x and y, respectively.

When they are, the z-coordinate is sometimes called the applicate. The words *abscissa, ordinate* and *applicate* are sometimes used to refer to coordinate axes rather than values.

Octant (Solid Geometry)

An octant in solid geometry is one of the eight divisions of a Euclidean three-dimensional coordinate system defined by the signs of the coordinates. It is similar to the two-dimensional quadrant and the one-dimensional ray. The generalisation of an octant is called orthant.

Numbering

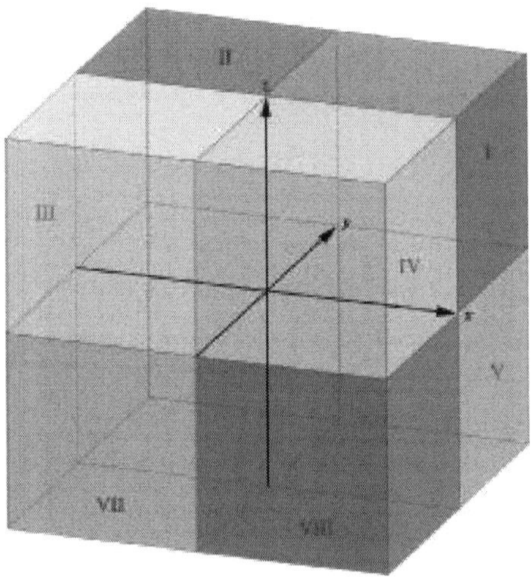

Figure: *For $z > 0$, the octants have the same numbers as the corresponding quadrants in the plane.*

Usually, the octant with all three positive coordinates is referred to as the *first octant*. There is no generally used naming convention for the other seven octants.

Number	*Name*	x	y	z	*Octal (+=0)*	*Octal (+=1)*
I	top-front-right	+	+	+	0	7
II	top-back-right	–	+	+	4	3
III	top-back-left	–	–	+	6	1
IV	top-front-left	+	–	+	2	5
V	bottom-front-right	+	+	–	1	6
VI	bottom-back-right	–	+	–	5	2
VII	bottom-back-left	–	–	–	7	0
VIII	bottom-front-left	+	–	–	3	4

Quadrant (Plane Geometry)

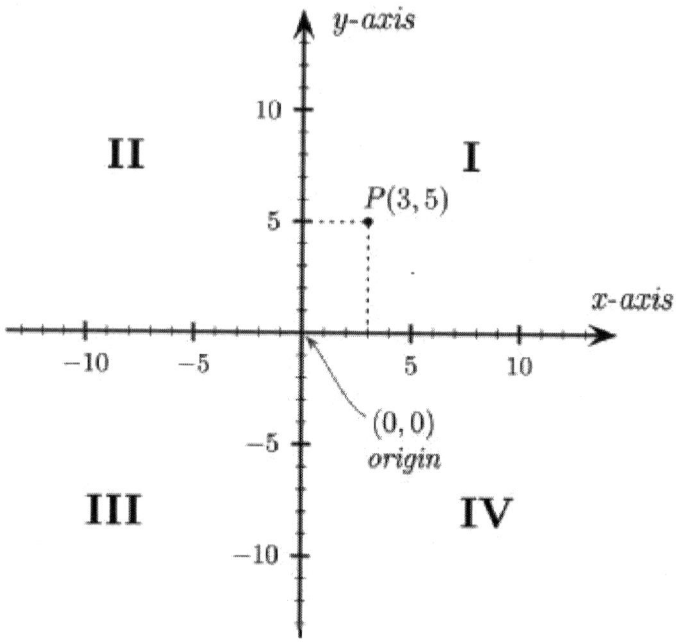

Figure: *The four quadrants of a Cartesian coordinate system.*

The axes of a two-dimensional Cartesian system divide the plane into four infinite regions, called quadrants, each bounded by two half-axes.

These are often numbered from 1st to 4th and denoted by Roman numerals: I (where the signs of the two coordinates are I (+,+), II (–,+), III (–,–), and IV (+,–). When the axes are drawn according to the mathematical custom, the numbering goes counter-clockwise starting from the upper right ("northeast") quadrant.

Cartesian Space

A Euclidean plane with a chosen Cartesian system is called a Cartesian plane.

Since Cartesian coordinates are unique and non-ambiguous, the points of a Cartesian plane can be identified with all possible pairs of real numbers; that is with the Cartesian product $\mathbb{R}^2 = \mathbb{R} \times \mathbb{R}$, where $\mathbb{R}$ is the set of all reals. In the same way one defines a Cartesian space of any dimension n, whose points can be identified with the tuples (lists) of n real numbers, that is, with $\mathbb{R}^n$.

Cartesian Formulas for the Plane

Distance Between Two Points

The Euclidean distance between two points of the plane with Cartesian coordinates (x_1, y_1) and (x_2, y_2) is

$$d = \sqrt{(x_2 - x_1)^2 + (y_2 - y_1)^2}.$$

This is the Cartesian version of Pythagoras' theorem. In three-dimensional space, the distance between points (x_1, y_1, z_1) and (x_2, y_2, z_2) is

$$d = \sqrt{(x_2 - x_1)^2 + (y_2 - y_1)^2 + (z_2 - z_1)^2}$$

which can be obtained by two consecutive applications of Pythagoras' theorem.

Euclidean Transformations

The Euclidean transformations or Euclidean motions are the (bijective) mappings of points of the Euclidean plane to themselves which preserve distances between points. There are four types of these mappings (also called isometries): translations, rotations, reflections and glide reflections.

Translation

Translating a set of points of the plane, preserving the distances and directions between them, is equivalent to adding a fixed pair of numbers (a, b) to the Cartesian coordinates of every point in the set. That is, if the original coordinates of a point are (x, y), after the translation they will be

$$(x', y') = (x + a, y + b).$$

Rotation

To rotate a figure counterclockwise around the origin by some angle θ is equivalent to replacing every point with coordinates (x, y) by the point with coordinates (x', y'), where

$$x' = x\cos\theta - y\sin\theta$$

$$y' = x\sin\theta + y\cos\theta.$$

Thus: $(x', y') = ((x\cos\theta - y\sin\theta), (x\sin\theta + y\cos\theta))$.

Reflection

If (x, y) are the Cartesian coordinates of a point, then $(-x, y)$ are the coordinates of its reflection across the second coordinate axis (the

Y axis), as if that line were a mirror. Likewise, $(x, -y)$ are the coordinates of its reflection across the first coordinate axis (the X axis). In more generality, reflection across a line through the origin making an angle θ with the x-axis, is equivalent to replacing every point with coordinates (x, y) by the point with coordinates (x', y'), where

$$x' = x\cos 2\theta + y\sin 2\theta$$

$$y' = x\sin 2\theta - y\cos 2\theta.$$

Thus: $(x', y') = ((x\cos 2\theta + y\sin 2\theta), (x\sin 2\theta - y\cos 2\theta))$.

Glide Reflection

A glide reflection is the composition of a reflection across a line followed by a translation in the direction of that line. It can be seen that the order of these operations does not matter (the translation can come first, followed by the reflection).

General Matrix Form of the Transformations

These Euclidean transformations of the plane can all be described in a uniform way by using matrices. The result (x', y') of applying a Euclidean transformation to a point (x, y) is given by the formula

$$(x', y') = (x, y)A + b$$

where A is a 2×2 orthogonal matrix and $b = (b_1, b_2)$ is an arbitrary ordered pair of numbers; that is,

$$x' = xA_{11} + yA_{21} + b_1$$

$$y' = xA_{12} + yA_{22} + b_2,$$

where

$A = \begin{pmatrix} A_{11} & A_{12} \\ A_{21} & A_{22} \end{pmatrix}$. [Note the use of row vectors for point coordinates and that the matrix is written on the right.]

To be *orthogonal*, the matrix A must have orthogonal rows with same Euclidean length of one, that is,

$$A_{11}A_{21} + A_{12}A_{22} = 0$$

and

$$A_{11}^2 + A_{12}^2 = A_{21}^2 + A_{22}^2 = 1.$$

This is equivalent to saying that A times its transpose must be the identity matrix. If these conditions do not hold, the formula describes a more general affine transformation of the plane provided that the determinant of A is not zero. The formula defines a translation if and only if A is the identity matrix. The transformation is a rotation around some point if and only if A is a rotation matrix, meaning that

$$A_{11}A_{22} - A_{21}A_{12} = 1.$$

A reflection or glide reflection is obtained when,

$$A_{11}A_{22} - A_{21}A_{12} = -1.$$

Assuming that translation is not used transformations can be combined by simply multiplying the associated transformation matrices.

Affine Transformation

Another way to represent coordinate transformations in Cartesian coordinates is through affine transformations. In affine transformations an extra dimension is added and all points are given a value of 1 for this extra dimension.

The advantage of doing this is that then all of the euclidean transformations become linear transformations and can be represented using matrix multiplication. The affine transformation is given by:

$$\begin{pmatrix} A_{11} & A_{21} & b_1 \\ A_{12} & A_{22} & b_2 \\ 0 & 0 & 1 \end{pmatrix} \begin{pmatrix} x \\ y \\ 1 \end{pmatrix} = \begin{pmatrix} x' \\ y' \\ 1 \end{pmatrix}.$$

[Note the A matrix from above was transposed. The matrix is on the left and column vectors for point coordinates are used.]

Using affine transformations multiple different euclidean transformations including translation can be combined by simply multiplying the corresponding matrices.

Scaling

An example of an affine transformation which is not a Euclidean motion is given by scaling. To make a figure larger or smaller is equivalent to multiplying the Cartesian coordinates of every point by the same positive number m. If (x, y) are the coordinates of a point on the original figure, the corresponding point on the scaled figure has coordinates

$$(x', y') = (mx, my).$$

If m is greater than 1, the figure becomes larger; if m is between 0 and 1, it becomes smaller.

Shearing

A shearing transformation will push the top of a square sideways to form a parallelogram. Horizontal shearing is defined by:

$$(x', y') = (x + ys, y)$$

Shearing can also be applied vertically:

$$(x', y') = (x, xs + y)$$

Orientation (Vector Space)

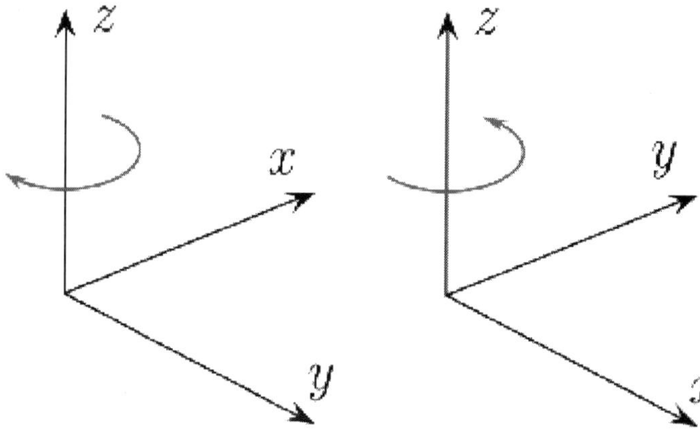

Figure: *The left-handed orientation is shown on the left, and the right-handed on the right.*

In mathematics, orientation is a geometric notion that in two dimensions allows one to say when a cycle goes around clockwise or counterclockwise, and in three dimensions when a figure is left-handed or right-handed.

In linear algebra, the notion of orientation makes sense in arbitrary dimensions. In this setting, the orientation of an ordered basis is a kind of asymmetry that makes a reflection impossible to replicate by means of a simple rotation.

Thus, in three dimensions, it is impossible to make the left hand of a human figure into the right hand of the figure by applying a rotation alone, but it is possible to do so by reflecting the figure in a mirror. As a result, in the three-dimensional Euclidean space, the two possible basis orientations are called right-handed and left-handed (or right-chiral and left-chiral).

The orientation on a real vector space is the arbitrary choice of which ordered bases are "positively" oriented and which are "negatively" oriented. In the three-dimensional Euclidean space, right-handed bases are typically declared to be positively oriented, but the choice is arbitrary, as they may also be assigned a negative orientation. A vector space with an orientation is called an oriented vector space, while one without a choice of orientation is called unoriented.

Definition

Let V be a finite-dimensional real vector space and let b_1 and b_2 be two ordered bases for V. It is a standard result in linear algebra that there exists a unique linear transformation $A : V \rightarrow V$ that takes b_1 to b_2.

The bases b_1 and b_2 are said to have the *same orientation* (or be consistently oriented) if A has positive determinant; otherwise they have *opposite orientations*. The property of having the same orientation defines an equivalence relation on the set of all ordered bases for V. If V is non-zero, there are precisely two equivalence classes determined by this relation. An orientation on V is an assignment of +1 to one equivalence class and −1 to the other.

Every ordered basis lives in one equivalence class or another. Thus any choice of a privileged ordered basis for V determines an orientation: the orientation class of the privileged basis is declared to be positive. For example, the standard basis on R^n provides a standard orientation on R^n (in turn, the orientation of the standard basis depends on the orientation of the Cartesian coordinate system on which it is built). Any choice of a linear isomorphism between V and R^n will then provide an orientation on V.

The ordering of elements in a basis is crucial. Two bases with a different ordering will differ by some permutation. They will have the same/opposite orientations according to whether the signature of this permutation is ± 1. This is because the determinant of a permutation matrix is equal to the signature of the associated permutation.

Similarly, let A be a nonsingular linear mapping of vector space R^n to R^n. This mapping is orientation-preserving if its determinant is positive. For instance, in R^3 a rotation around the Z Cartesian axis by an angle α is orientation-preserving:

$$\mathbf{A}_1 = \begin{pmatrix} \cos\alpha & -\sin\alpha & 0 \\ \sin\alpha & \cos\alpha & 0 \\ 0 & 0 & 1 \end{pmatrix}$$

while a reflection by the XY Cartesian plane is not orientation-preserving:

$$\mathbf{A}_2 = \begin{pmatrix} 1 & 0 & 0 \\ 0 & 1 & 0 \\ 0 & 0 & -1 \end{pmatrix}$$

Zero-dimensional Case

The concept of orientation defined above did not quite apply to zero-dimensional vector spaces (as the only empty matrix is the identity (with determinant 1), so there will be only one equivalence class). However, it is useful to be able to assign different orientations to a point (e.g. orienting the boundary of a 1-dimensional manifold).

A more general definition of orientation that works regardless of dimension is the following: An orientation on V is a map from the set of ordered bases of V to the set $\{\pm 1\}$ that is invariant under base changes with positive determinant and changes sign under base changes with negative determinant (it is equivarient with respect to the homomorphism $GL_n \to \pm 1$).

The set of ordered bases of the zero-dimensional vector space has one element (the empty set), and so there are two maps from this set to $\{\pm 1\}$. A subtle point is that a zero-dimensional vector space is naturally (canonically) oriented, so we can talk about an orientation being positive (agreeing with the canonical orientation) or negative (disagreeing). An application is interpreting the Fundamental theorem of calculus as a special case of Stokes' theorem.

Two ways of seeing this are:

- A zero-dimensional vector space is a point, and there is a unique map from a point to a point, so every zero-dimensional vector space is naturally identified with R^0, and thus is oriented.
- The 0th exterior power of a vector space is the ground field, which here is R^1, which has an orientation (given by the standard basis).

Alternate Viewpoints

Multilinear Algebra

For any n-dimensional real vector space V we can form the kth-exterior power of V, denoted $\Lambda^k V$. This is a real vector space of dimension $\binom{n}{k}$. The vector space $\Lambda^n V$ (called the *top exterior power*) therefore has dimension 1. That is, $\Lambda^n V$ is just a real line. There is no *a priori* choice of which direction on this line is positive. An orientation is just such a choice. Any nonzero linear form ω on Ë^{n}V determines an orientation of $V$by declaring that x is in the positive direction when ω (x) > 0. To connect with the basis point of view we say that the positively oriented bases are those on which ω evaluates to a positive number (since ω is an n-form we can evaluate it on an ordered set of n vectors, giving an element of R). The form ω is called an orientation form. If $\{e_i\}$ is a privileged basis for $V$and $\{e_i^*\}$ is the dual basis, then the orientation form giving the standard orientation is $e_1^* \wedge e_2^* \wedge \ldots \wedge e_n^*$. The connection of this with the determinant point of view is: the determinant of an endomorphism $T : V \rightarrow V$ can be interpreted as the induced action on the top exterior power.

Orientation on Manifolds

One can also discuss orientation on manifolds. Each point p on an n-dimensional differentiable manifold has a tangent space T_pM which is an n-dimensional real vector space. One can assign to each of these vector spaces an orientation. However, one would like to know whether it is possible to choose the orientations so that they "vary smoothly" from point to point. Due to certain topological restrictions, there are situations when this is impossible. A manifold which admits a smooth choice of orientations for its tangents spaces is said to be *orientable*.

Axes Conventions

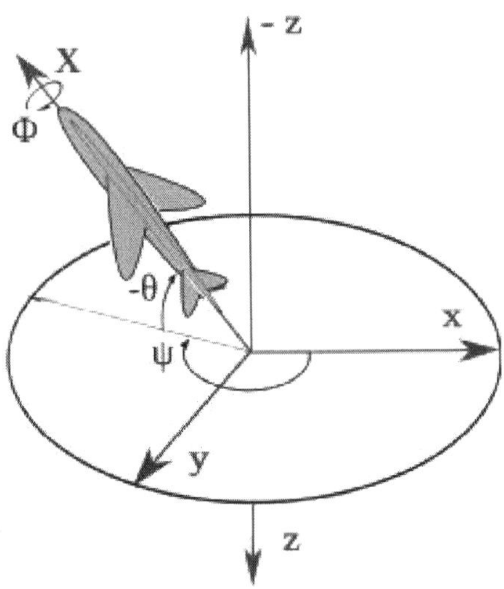

Figure: *Heading, elevation and bank angles (Z-Y'-X") for an aircraft. The aircraft's pitch and yaw axes Y and Z are not shown, and its fixed reference frame xyz has been shifted backwards from its centre of gravity (preserving angles) for clarity. Axes named according to the air norm DIN 9300*

Mobile objects are normally tracked from an external frame considered fixed. Other frames can be defined on those mobile objects to deal with relative positions for other objects. Finally, attitudes or orientations can be described by a relationship between the external frame and the one defined over the mobile object. The orientation of a vehicle is normally referred to as *attitude*. It is described normally by the orientation of a frame fixed in the body relative to a fixed reference frame. The attitude is described by *attitude coordinates*, and consists of at least three coordinates.

While from a geometrical point of view the different methods to describe orientations are defined using only some reference frames, in engineering applications it is important also to describe how these frames are attached to the lab and the body in motion.

Due to the special importance of international conventions in air vehicles, several organisations have published standards to be followed. For example, German DIN has published the DIN 9300 norm for aircraft (adopted by ISO as ISO 1151–2:1985).

Ground Reference Frames: ENU and NED

Basically, as lab frame or reference frame, there are two kinds of conventions for the frames (sometimes named LVLH, local vertical, local horizontal):

- East, North, Up, referred as ENU
- North, East, Down, referred as NED, used specially in aerospace

  These frames are location dependent. For movements around the globe, like air or sea navigation, the frames are defined as tangent to the lines of coordinates.

- East-West tangent to parallels,
- North-South tangent to meridians, and
- Up-Down in the direction to the centre of the earth (when using a spherical Earth simplification), or in the direction normal to the local tangent plane (using an oblate spheroidal or geodetic ellipsoidal model of the earth) which does not generally pass through the centre of the Earth.

To establish a standard convention to describe attitudes, it is required to establish at least the axes of the reference system and the axes of the rigid body or vehicle. When an ambiguous notation system is used (such as Euler angles) also the used convention should be stated. Nevertheless most used notations (matrices and quaternions) are unambiguous.

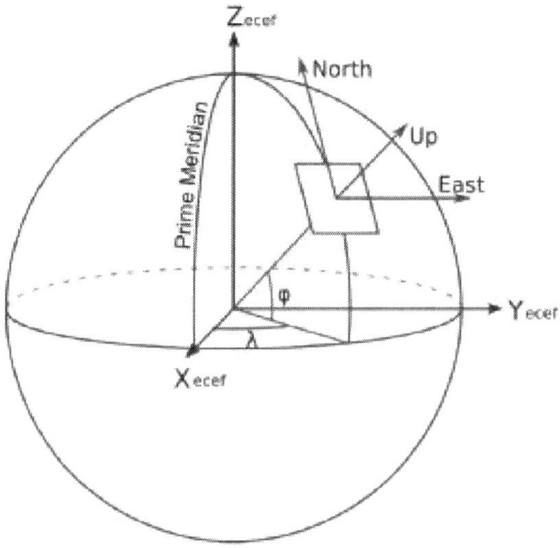

Figure: *Earth Centred Earth Fixed and East, North, Up coordinates.*

Tait–Bryan angles are often used to describe a vehicle's attitude with respect to a chosen reference frame, though any other notation can be used. The positive *x*-axis in vehicles points always in the direction of movement. For positive *y*- and *z*-axis, we have to face two different conventions:

- In case of land vehicles like cars, tanks etc., which use the ENU-system (East-North-Up) as external reference (*world frame*), the vehicle's positive *y*- or pitch axis always points to its left, and the positive *z*- or yaw axis always points up.
- By contrast, in case of air and sea vehicles like submarines, ships, airplanes etc., which use the NED-system (North-East-Down) as external reference (*world frame*), the vehicle's positive *y*- or pitch axis always points to its right, and its positive *z*- or yaw axis always points down.
- Finally, in case of space vehicles like space shuttles etc., a modification of the latter convention is used, where the vehicle's positive *y*- or pitch axis again always points to its right, and its positive z- or yaw axis always points down, but "down" now may have two different meanings: If a so-called *local frame* is used as external reference, its positive z-axis points "down" to the centre of the earth as it does in case of the earlier mentioned NED-system, but if the *inertial frame* is used as reference, its positive z-axis will point now to the North Celestial Pole, and its positive x-axis to the Vernal Equinox or some other reference meridian.

Frames Mounted on Vehicles

Specially for aircraft, these frames do not need to agree with the earth-bound frames in the up-down line. It must be agreed what ENU and NED mean in this context.

Conventions for Land Vehicles

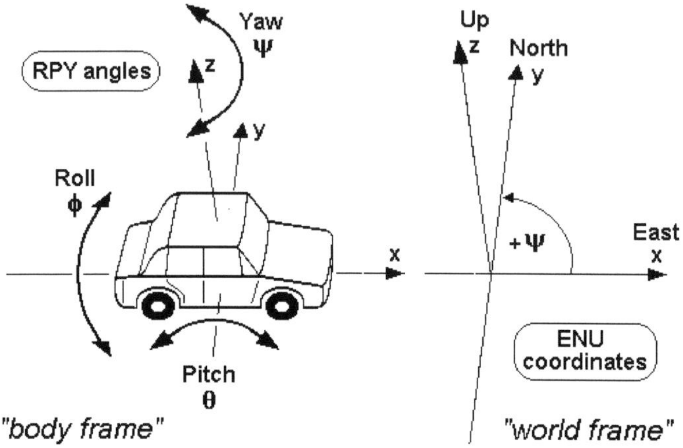

Figure: *RPY angles of cars and other land vehicles*

For land vehicles is rare to describe their complete orientation, except when speaking about electronic stability control or satellite navigation. In this case, the convention is normally the one of the adjacent drawing, where RPY stands for roll-pitch-yaw.

Conventions for Sea Vehicles

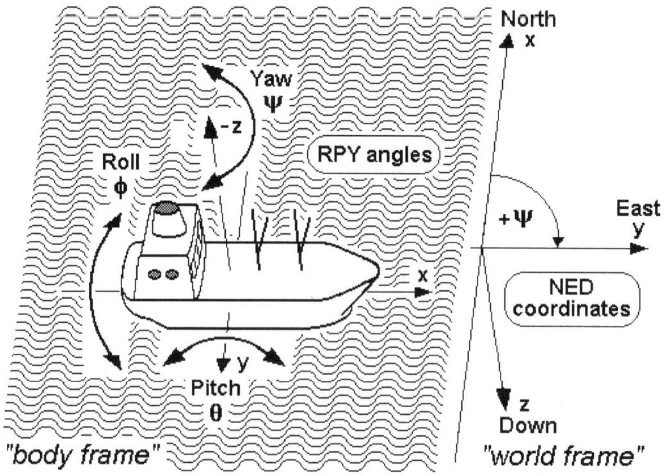

Figure: *RPY angles of ships and other sea vehicles*

As well as aircraft, the same terminology is used for the motion of ships and boats. It is interesting to note that some words commonly used were introduced in maritime navigation. For example, the *yaw* angle or heading, has a nautical origin, with the meaning of "bending out of the course". Etymologically, it is related with the verb 'to go'. It is related to the concept of bearing. It is typically assigned the shorthand notation Ψ.

Conventions for Aircraft Local Reference Frames

Coordinates to describe an aircraft attitude (Heading, Elevation and Bank) are normally given relative to a reference control frame located in a control tower, and therefore ENU, relative to the position of the control tower on the earth surface.

Coordinates to describe observations made from an aircraft are normally given relative to its intrinsic axes, but normally using as positive the coordinate pointing downwards, where the interesting points are located. Therefore they are normally NED.

These axes are normally taken so that X axis is the longitudinal axis pointing ahead, Z axis is the vertical axis pointing downwards, and the Y axis is the lateral one, pointing in such a way that the frame is right handed.

The motion of an aircraft is often described in terms of rotation about these axes, so rotation about the X-axis is called rolling, rotation about the Y-axis is called pitching, and rotation about the Z-axis is called yawing.

Frames for Space Navigation

For satellites orbiting the earth it is normal to use the Equatorial coordinate system. The projection of the Earth's equator onto the celestial sphere is called the celestial equator. Similarly, the projections of the Earth's north and south geographic poles become the north and south celestial poles, respectively. Deep space satellites use other Celestial coordinate system, like the Ecliptic coordinate system.

Conventions for Space Ships as Aircraft

If the goal is to keep the shuttle during its orbits in a constant attitude with respect to the sky, e.g. in order to perform certain astronomical observations, the preferred reference is the *inertial frame*, and the RPY angle vector (0 | 0 | 0) describes an attitude then, where the shuttle's wings are kept permanently parallel to the earth's equator, its nose points permanently to the vernal equinox, and its belly towards the Northern polar star. (Note that rockets and missiles more commonly follow the conventions for aircraft where the RPY angle vector (0 | 0 | 0) points north, rather than towards the vernal equinox).

On the other hand, if it's the goal to keep the shuttle during its orbits in an constant attitude with respect to the surface of the earth, the preferred reference will be the *local frame*, with the RPY angle vector (0 | 0 | 0) describing an attitude, where the shuttle's wings are parallel to the earth's surface, its nose points to its heading, and its belly down towards the centre of the earth.

Frames Used to Describe Attitudes

Normally the frames used to describe a vehicle's local observations are the same frames used to describe its attitude respect the ground tracking stations.

An important case in which this does not apply are aircraft. Aircraft observations are performed downwards and therefore normally NED axes convention applies.

Nevertheless, when attitudes respect ground stations are given, a relationship between the local earth-bound frame and the onboard ENU frame is used.

Right-hand Rule

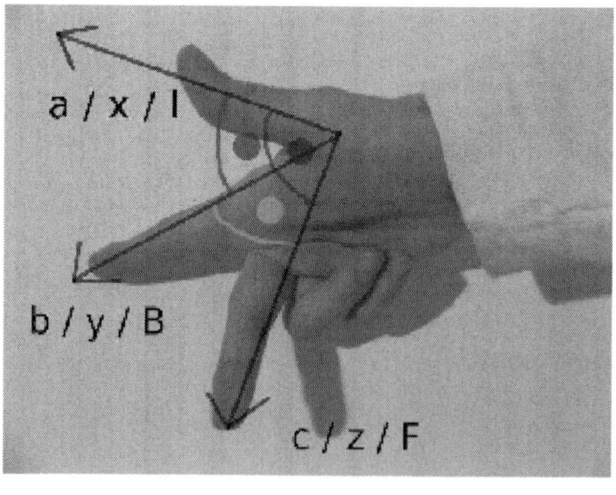

Figure: *Use of right hand*

In mathematics and physics, the right-hand rule is a common mnemonic for understanding notation conventions for vectors in 3 dimensions. It was invented for use in electromagnetism by British physicist John Ambrose Fleming in the late 19th century.

When choosing three vectors that must be at right angles to each other, there are two distinct solutions, so when expressing this idea in mathematics, one must remove the ambiguity of which solution is meant.

There are variations on the mnemonic depending on context, but all variations are related to the one idea of choosing a convention.

Direction Associated with an Ordered Pair of Directions

One form of the right-hand rule is used in situations in which an ordered operation must be performed on two vectors a and b that has a result which is a vector c perpendicular to both a and b.

The most common example is the vector cross product. The right-hand rule imposes the following procedure for choosing one of the two directions.

$$\vec{a}\times\vec{b}=\vec{c}$$

- With the thumb, index, and middle fingers at right angles to each other (with the index finger pointed straight), the middle finger points in the direction of c when the thumb represents a and the index finger represents b.

Other (equivalent) finger assignments are possible. For example, the first (index) finger can represent a, the first vector in the product; the second (middle) finger, b, the second vector; and the thumb, c, the product.

Direction Associated with a Rotation

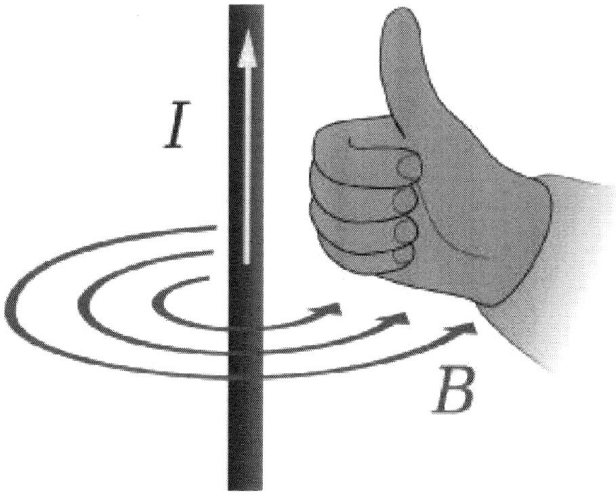

Figure: *Prediction of direction of field (B), given that the current I flows in the direction of the thumb*

Figure: *The right-hand rule as applied to motion produced with screw threads*

A different form of the right-hand rule, sometimes called the *right-hand grip rule* or the *corkscrew-rule* or the *right-hand thumb rule*, is used in situations where a vector must be assigned to the *rotation* of a body, a magnetic field or a fluid. Alternatively, when a rotation is specified by a vector, and it is necessary to understand the way in which the rotation occurs, the right-hand grip rule is applicable.

This version of the rule is used in two complementary applications of Ampère's circuital law:

1. An electric current passes through a solenoid, resulting in a magnetic field. When you wrap your right hand around the solenoid with your fingers in the direction of the conventional current, your thumb points in the direction of the magnetic north pole.
2. An electric current passes through a straight wire. Here, the thumb points in the direction of the conventional current (from positive to negative), and the fingers point in the direction of the magnetic lines of flux.

The principle is also used to determine the direction of the torque vector. If you grip the imaginary axis of rotation of the rotational force so that your fingers point in the direction of the force, then the extended thumb points in the direction of the torque vector. The right-hand grip rule is a convention derived from the right-hand rule convention for vectors. When applying the rule to current in a straight wire for example, the direction of the magnetic field (counterclockwise instead of clockwise when viewed from the tip of the thumb) is a result of this convention and not an underlying physical phenomenon.

Applications

The first form of the rule is used to determine the direction of the cross product of two vectors. This leads to widespread use in physics, wherever the cross product occurs. A list of physical quantities whose directions are related by the right-hand rule is given below. (Some of these are related only indirectly to cross products, and use the second form.)

- The angular velocity of a rotating object and the rotational velocity of any point on the object
- A torque, the force that causes it, and the position of the point of application of the force
- A magnetic field, the position of the point where it is determined, and the electric current (or change in electric flux) that causes it
- A magnetic field in a coil of wire and the electric current in the wire
- The force of a magnetic field on a charged particle, the magnetic field itself, and the velocity of the object
- The vorticity at any point in the field of flow of a fluid
- The induced current from motion in a magnetic field (known as Fleming's right-hand rule)
- The x, y and z unit vectors in a Cartesian coordinate system can be chosen to follow the right-hand rule. Right-handed coordinate systems are often used in rigid body physics and kinematics.

Fleming's left-hand rule is a rule for finding the direction of the thrust on a conductor carrying a current in a magnetic field.

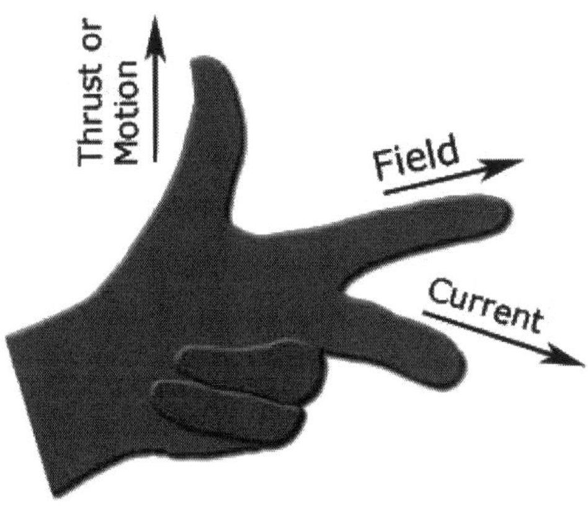

Figure: *Fleming's left-hand rule.*

Left-hand Rule

In certain situations, it may be useful to use the opposite convention, where one of the vectors is reversed and so creates a left-handed triad instead of a right-handed triad. An example of this situation is for left-handed materials. Normally, for an electromagnetic wave, the electric and magnetic fields, and the direction of propagation of the wave obey the right-hand rule. However, left-handed materials have special properties, notably the negative refractive index. It makes the direction of propagation point in the opposite direction.

De Graaf's translation of Fleming's left-hand rule - which uses thrust, field and current - and the right-hand rule, is the FBI rule. The FBI rule changes thrust into F (Lorentz force), B (direction of the magnetic field) and I (current). The FBI rule is easily remembered by US citizens because of the commonly known abbreviation for the Federal Bureau of Investigation.

Symmetry

Vector	*Right-hand*	*Right-hand*	*Right-hand*	*Left-hand*	*Left-hand*	*Left-Hand*
a, x or I	Thumb	Fingers or palm	First or Index	Thumb	Fingers or palm	First or index
b, y or B	First or index	Thumb	Fingers or palm	Fingers or palm	First or index	Thumb
c, z or F	Fingers or palm	First or index	Thumb	First or index	Thumb	Fingers or palm

In Two Dimensions

Fixing or choosing the x-axis determines the y-axis up to direction. Namely, the y-axis is necessarily the perpendicular to the x-axis through the point marked 0 on the x-axis. But there is a choice of which of the two half lines on the perpendicular to designate as positive and which as negative. Each of these two choices determines a different orientation (also called *handedness*) of the Cartesian plane.

The usual way of orienting the axes, with the positive x-axis pointing right and the positive y-axis pointing up (and the x-axis being the "first" and the y-axis the "second" axis) is considered the *positive* or *standard* orientation, also called the *right-handed* orientation.

A commonly used mnemonic for defining the positive orientation is the *right hand rule*. Placing a somewhat closed right hand on the plane with the thumb pointing up, the fingers point from the x-axis to the y-axis, in a positively oriented coordinate system. The other way of orienting the axes is following the *left hand rule*, placing the left hand on the plane with the thumb pointing up.

When pointing the thumb away from the origin along an axis, the curvature of the fingers indicates a positive rotation along that axis. Regardless of the rule used to orient the axes, rotating the coordinate system will preserve the orientation. Switching any two axes will reverse the orientation.

In Three Dimensions

Once the x- and y-axes are specified, they determine the line along which the z-axis should lie, but there are two possible directions on this line. The two possible coordinate systems which result are called 'right-handed' and 'left-handed'. The standard orientation, where the xy-plane is horizontal and the z-axis points up (and the x- and the y-axis form a positively oriented two-dimensional coordinate system in the xy-plane if observed from *above* the xy-plane) is called right-handed or positive. The name derives from the right-hand rule. If the index finger of the right hand is pointed forward, the middle finger bent inward at a right angle to it, and the thumb placed at a right angle to both, the three fingers indicate the relative directions of the x-, y-, and z-axes in a *right-handed* system. The thumb indicates the x-axis, the index finger the y-axis and the middle finger the z-axis. Conversely, if the same is done with the left hand, a left-handed system results. Figure depicts a left and a right-handed coordinate system. Because a three-dimensional object is represented on the two-dimensional screen, distortion and ambiguity result. The axis pointing downward (and to the right) is also meant to point *towards* the observer, whereas the "middle" axis is meant to point *away* from the observer. The red circle is *parallel* to the horizontal xy-plane and indicates rotation from the x-axis to the y-axis (in both cases). Hence the red arrow passes *in front of* the z-axis. Figure is another attempt at depicting a right-handed coordinate system. Again, there is an ambiguity caused by projecting the three-dimensional coordinate system into the plane. This corresponds to the two possible orientations of the coordinate system. Thus the "correct" way to view is to imagine the x-axis as pointing *towards* the observer and thus seeing a concave corner.

Representing a Vector in the Standard Basis

A point in space in a Cartesian coordinate system may also be represented by a position vector, which can be thought of as an arrow pointing from the origin of the coordinate system to the point. If the coordinates represent spatial positions (displacements), it is common to represent the vector from the origin to the point of interest as r.

In two dimensions, the vector from the origin to the point with Cartesian coordinates (x, y) can be written as:

$$r = x\mathrm{i} + y\mathrm{j}$$

where i = (1, 0), and j = (0, 1)are unit vectors in the direction of the x-axis and y-axis respectively, generally referred to as the *standard basis* (in some application areas these may also be referred to as versors). Similarly, in three dimensions, the vector from the origin to the point with Cartesian coordinates (x,y,z) can be written as:

$$r = xi + yj + zk$$

where k = (0, 0, 1) is the unit vector in the direction of the z-axis.

There is no *natural* interpretation of multiplying vectors to obtain another vector that works in all dimensions, however there is a way to use complex numbers to provide such a multiplication. In a two dimensional cartesian plane, identify the point with coordinates (x, y) with the complex number $z = x + iy$. Here, i is the complex number whose square is the real number –1 and is identified with the point with coordinates (0,1), so it is not the unit vector in the direction of the x-axis (this confusion is just an unfortunate historical accident). Since the complex numbers can be multiplied giving another complex number, this identification provides a means to "multiply" vectors. In a three dimensional cartesian space a similar identification can be made with a subset of the quaternions.

Chapter

2

Polar Coordinate System

In mathematics, the polar coordinate system is a two-dimensional coordinate system in which each point on a plane is determined by a distance from a fixed point and an angle from a fixed direction.

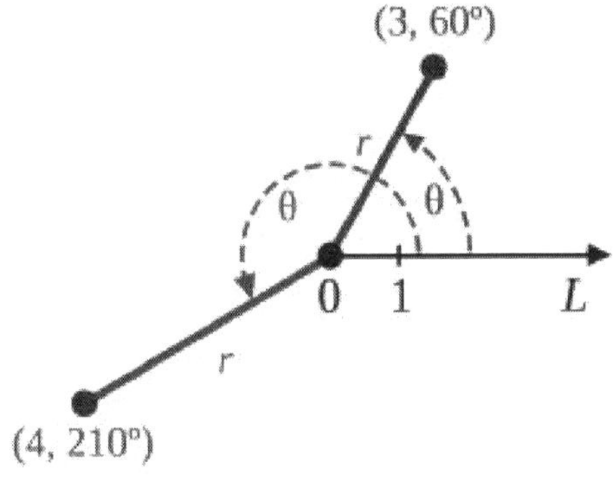

***Figure:** Points in the polar coordinate system with pole O and polar axis L. In green, the point with radial coordinate 3 and angular coordinate 60 degrees, or (3,60°). In blue, the point (4,210°).*

The fixed point (analogous to the origin of a Cartesian system) is called the *pole*, and the ray from the pole in the fixed direction is the *polar axis*. The distance from the pole is called the *radial coordinate* or *radius*, and the angle is the *angular coordinate*, *polar angle*, or *azimuth*.

History

The concepts of angle and radius were already used by ancient peoples of the 1st millennium BCE. The Greek astronomer and astrologer

Hipparchus (190–120 BCE) created a table of chord functions giving the length of the chord for each angle, and there are references to his using polar coordinates in establishing stellar positions. In *On Spirals*, Archimedes describes the Archimedean spiral, a function whose radius depends on the angle. The Greek work, however, did not extend to a full coordinate system.

From the 8th century CE onward, astronomers developed methods for approximating and calculating the direction to Makkah (qibla)—and its distance—from any location on the Earth. From the 9th century onward they were using spherical trigonometry and map projection methods to determine these quantities accurately. The calculation is essentially the conversion of the equatorial polar coordinates of Mecca (i.e. its longitude and latitude) to its polar coordinates (i.e. its qibla and distance) relative to a system whose reference meridian is the great circle through the given location and the Earth's poles, and whose polar axis is the line through the location and its antipodal point.

There are various accounts of the introduction of polar coordinates as part of a formal coordinate system. The full history of the subject is described in Harvard professor Julian Lowell Coolidge's *Origin of Polar Coordinates.* Grégoire de Saint-Vincent and Bonaventura Cavalieri independently introduced the concepts in the mid-seventeenth century. Saint-Vincent wrote about them privately in 1625 and published his work in 1647, while Cavalieri published his in 1635 with a corrected version appearing in 1653. Cavalieri first used polar coordinates to solve a problem relating to the area within an Archimedean spiral. Blaise Pascal subsequently used polar coordinates to calculate the length of parabolic arcs.

In *Method of Fluxions* (written 1671, published 1736), Sir Isaac Newton examined the transformations between polar coordinates, which he referred to as the "Seventh Manner; For Spirals", and nine other coordinate systems. In the journal *Acta Eruditorum* (1691), Jacob Bernoulli used a system with a point on a line, called the *pole* and *polar axis* respectively. Coordinates were specified by the distance from the pole and the angle from the *polar axis*. Bernoulli's work extended to finding the radius of curvature of curves expressed in these coordinates.

The actual term *polar coordinates* has been attributed to Gregorio Fontana and was used by 18th-century Italian writers. The term appeared in English in George Peacock's 1816 translation of Lacroix's

Differential and Integral Calculus. Alexis Clairaut was the first to think of polar coordinates in three dimensions, and Leonhard Euler was the first to actually develop them.

Conventions

The radial coordinate is often denoted by r, and the angular coordinate by θ or t.

Angles in polar notation are generally expressed in either degrees or radians (2π rad being equal to 360°). Degrees are traditionally used in navigation, surveying, and many applied disciplines, while radians are more common in mathematics and mathematical physics. In many contexts, a positive angular coordinate means that the angle θ is measured counterclockwise from the axis. In mathematical literature, the polar axis is often drawn horizontal and pointing to the right.

Uniqueness of Polar Coordinates

Adding any number of full turns (360°) to the angular coordinate does not change the corresponding direction. Also, a negative radial coordinate is best interpreted as the corresponding positive distance measured in the opposite direction. Therefore, the same point can be expressed with an infinite number of different polar coordinates $(r, \theta \pm n \times 360°)$ or $(-r, \theta \pm (2n + 1)180°)$, where n is any integer. Moreover, the pole itself can be expressed as $(0, \theta)$ for any angle θ. Where a unique representation is needed for any point, it is usual to limit r to non-negative numbers ($r \geq 0$) and θ to the interval [0, 360°) or (−180°, 180°] (in radians, $[0, 2\pi)$ or $(-\pi, \pi]$). One must also choose a unique azimuth for the pole, e.g., $\theta = 0$.

Polar Equation of a Curve

The equation defining an algebraic curve expressed in polar coordinates is known as a *polar equation*. In many cases, such an equation can simply be specified by defining r as a function of θ. The resulting curve then consists of points of the form $(r(\theta), \theta)$ and can be regarded as the graph of the polar function r.

Different forms of symmetry can be deduced from the equation of a polar function r. If $r(-\theta) = r(\theta)$ the curve will be symmetrical about the horizontal (0°/180°) ray, if $r(\pi - \theta) = r(\theta)$ it will be symmetric about the vertical (90°/270°) ray, and if $r(\theta - \alpha) = r(\theta)$ it will be rotationally symmetric α counterclockwise about the pole.

Because of the circular nature of the polar coordinate system, many curves can be described by a rather simple polar equation, whereas their Cartesian form is much more intricate. Among the best known of these curves are the polar rose, Archimedean spiral, lemniscate, limaçon, and cardioid. For the circle, line, and polar rose below, it is understood that there are no restrictions on the domain and range of the curve.

Circle

The general equation for a circle with a centre at (r_0, φ) and radius a is

$$r^2 - 2rr_0 \cos(\theta - \varphi) + r_0^2 = a^2.$$

This can be simplified in various ways, to conform to more specific cases, such as the equation

$$r(\theta) = a$$

for a circle with a centre at the pole and radius a.

When $r_0 = a$, or when the origin lies on the circle, the equation becomes

$$r = 2a\cos(\theta - \varphi).$$

In the general case, the equation can be solved for r, giving

$$r = r_0 \cos(\theta - \varphi) + \sqrt{a^2 - r_0^2 \sin^2(\theta - \varphi)}$$

the solution with a minus sign in front of the square root giving the same curve.

Line

Radial lines (those running through the pole) are represented by the equation

$$\theta = \varphi$$

where φ is the angle of elevation of the line; that is, $\varphi = \arctan m$ where m is the slope of the line in the Cartesian coordinate system. The non-radial line that crosses the radial line $\theta = \varphi$ perpendicularly at the point (r_0, φ) has the equation

$$r(\theta) = r_0 \sec(\theta - \varphi).$$

Otherwise stated (r_0, φ) is the point in which the tangent intersects the imaginary circle of radius r_0.

Polar Rose

A polar rose is a famous mathematical curve that looks like a petaled flower, and that can be expressed as a simple polar equation,

$$r(\theta) = a\cos(k\theta + \varphi_0)$$

for any constant φ_0 (including 0). If k is an integer, these equations will produce a k-petaled rose if k is odd, or a $2k$-petaled rose if k is even.

If k is rational but not an integer, a rose-like shape may form but with overlapping petals. Note that these equations never define a rose with 2, 6, 10, 14, etc. petals. The variable a represents the length of the petals of the rose.

Archimedean Spiral

The Archimedean spiral is a famous spiral that was discovered by Archimedes, which also can be expressed as a simple polar equation. It is represented by the equation

$$r(\theta) = a + b\theta.$$

Changing the parametre a will turn the spiral, while b controls the distance between the arms, which for a given spiral is always constant.

The Archimedean spiral has two arms, one for $\theta > 0$ and one for $\theta < 0$. The two arms are smoothly connected at the pole. Taking the mirror image of one arm across the 90°/270° line will yield the other arm.

This curve is notable as one of the first curves, after the conic sections, to be described in a mathematical treatise, and as being a prime example of a curve that is best defined by a polar equation.

Complex Numbers

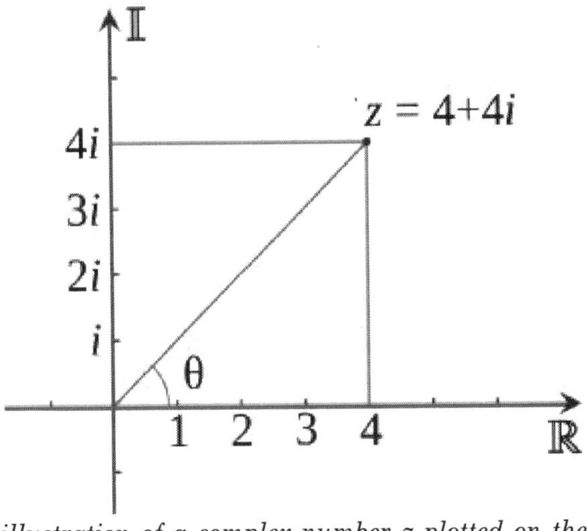

Figure: *An illustration of a complex number z plotted on the complex plane*

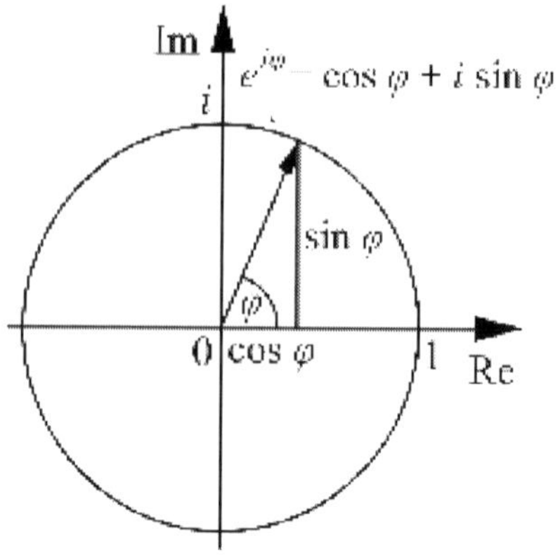

Figure: *An illustration of a complex number plotted on the complex plane using Euler's formula*

Every complex number can be represented as a point in the complex plane, and can therefore be expressed by specifying either the point's Cartesian coordinates (called rectangular or Cartesian form) or the point's polar coordinates (called polar form). The complex number z can be represented in rectangular form as

$$z = x + iy$$

where i is the imaginary unit, or can alternatively be written in polar form (via the conversion formulae given above) as

$$z = r\cdot(\cos\theta + i\sin\theta)$$

and from there as

$$z = re^{i\theta}$$

where e is Euler's number, which are equivalent as shown by Euler's formula. (Note that this formula, like all those involving exponentials of angles, assumes that the angle θ is expressed in radians.) To convert between the rectangular and polar forms of a complex number, the conversion formulae given above can be used.

For the operations of multiplication, division, and exponentiation of complex numbers, it is generally much simpler to work with complex numbers expressed in polar form rather than rectangular form. From the laws of exponentiation:

- Multiplication: $r_0e^{i\theta_0} \cdot r_1e^{i\theta_1} = r_0r_1e^{i(\theta_0+\theta_1)}$
- Division: $\dfrac{r_0e^{i\theta_0}}{r_1e^{i\theta_1}} = \dfrac{r_0}{r_1}e^{i(\theta_0-\theta_1)}$
- Exponentiation (De Moivre's formula): $(re^{i\theta})^n = r^ne^{in\theta}$

Connection to Spherical and Cylindrical Coordinates

The polar coordinate system is extended into three dimensions with two different coordinate systems, the cylindrical and spherical coordinate system.

Applications

Polar coordinates are two-dimensional and thus they can be used only where point positions lie on a single two-dimensional plane. They are most appropriate in any context where the phenomenon being considered is inherently tied to direction and length from a centre point.

For instance, the examples above show how elementary polar equations suffice to define curves—such as the Archimedean spiral—whose equation in the Cartesian coordinate system would be much more intricate.

Moreover, many physical systems—such as those concerned with bodies moving around a central point or with phenomena originating from a central point—are simpler and more intuitive to model using polar coordinates. The initial motivation for the introduction of the polar system was the study of circular and orbital motion.

Position and Navigation

Polar coordinates are used often in navigation, as the destination or direction of travel can be given as an angle and distance from the object being considered.

For instance, aircraft use a slightly modified version of the polar coordinates for navigation. In this system, the one generally used for any sort of navigation, the 0° ray is generally called heading 360, and the angles continue in a clockwise direction, rather than counterclockwise, as in the mathematical system.

Heading 360 corresponds to magnetic north, while headings 90, 180, and 270 correspond to magnetic east, south, and west, respectively. Thus, an aircraft travelling 5 nautical miles due east will be travelling 5 units at heading 90 (read zero-niner-zero by air traffic control).

Modelling

Systems displaying radial symmetry provide natural settings for the polar coordinate system, with the central point acting as the pole. A prime example of this usage is the groundwater flow equation when applied to radially symmetric wells. Systems with a radial force are also good candidates for the use of the polar coordinate system. These systems include gravitational fields, which obey the inverse-square law, as well as systems with point sources, such as radio antennas.

Radially asymmetric systems may also be modeled with polar coordinates.

For example, a microphone's pickup pattern illustrates its proportional response to an incoming sound from a given direction, and these patterns can be represented as polar curves. The curve for a standard cardioid microphone, the most common unidirectional microphone, can be represented as $r = 0.5 + 0.5\sin(\theta)$ at its target design frequency. The pattern shifts towards omnidirectionality at lower frequencies.

Cylindrical Coordinate System

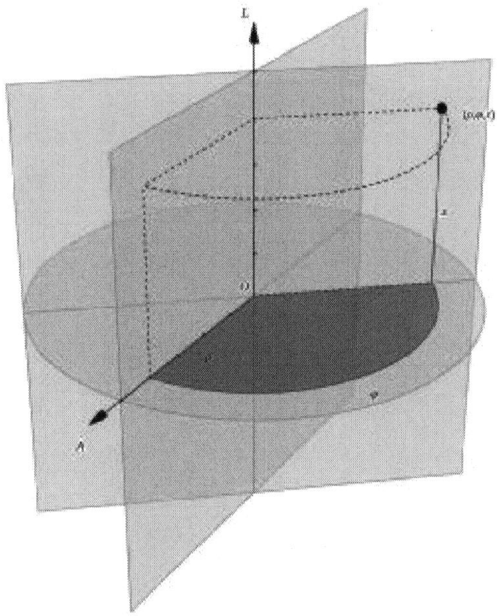

Figure: *A cylindrical coordinate system with origin O, polar axis A, and longitudinal axis L. The dot is the point with radial distance* $\rho = 4$*, angular coordinate* $\varphi = 130°$*, and height* $z = 4$*.*

A cylindrical coordinate system is a three-dimensional coordinate system that specifies point positions by the distance from a chosen reference axis, the direction from the axis relative to a chosen reference direction, and the distance from a chosen reference plane perpendicular to the axis. The latter distance is given as a positive or negative number depending on which side of the reference plane faces the point.

The *origin* of the system is the point where all three coordinates can be given as zero. This is the intersection between the reference plane and the axis.

The axis is variously called the *cylindrical* or *longitudinal* axis, to differentiate it from the *polar axis*, which is the ray that lies in the reference plane, starting at the origin and pointing in the reference direction.

The distance from the axis may be called the *radial distance* or *radius*, while the angular coordinate is sometimes referred to as the *angular position* or as the *azimuth*.

The radius and the azimuth are together called the *polar coordinates*, as they correspond to a two-dimensional polar coordinate system in the plane through the point, parallel to the reference plane.

The third coordinate may be called the *height* or *altitude* (if the reference plane is considered horizontal), *longitudinal position*, or *axial position*. Cylindrical coordinates are useful in connection with objects and phenomena that have some rotational symmetry about the longitudinal axis, such as water flow in a straight pipe with round cross-section, heat distribution in a metal cylinder, electromagnetic fields produced by an electric current in a long, straight wire, and so on.

It is sometimes called "cylindrical polar coordinate" and "polar cylindrical coordinate", and is sometime used to specify the position of stars in a galaxy ("galactocentric cylindrical polar coordinate").

Definition

The three coordinates (ρ, φ, z) of a point P are defined as:

- The radial distance ρ is the Euclidean distance from the z axis to the point P.
- The azimuth φ is the angle between the reference direction on the chosen plane and the line from the origin to the projection of P on the plane.
- The height z is the signed distance from the chosen plane to the point P.

Unique Cylindrical Coordinates

As in polar coordinates, the same point with cylindrical coordinates (ρ, φ, z) has infinitely many equivalent coordinates, namely (ρ, $\varphi \pm n \times 360°$, z) and ($-\rho$, $\varphi \pm (2n + 1) \times 180°$, z), where n is any integer. Moreover, if the radius ρ is zero, the azimuth is arbitrary.

In situations where one needs a unique set of coordinates for each point, one may restrict the radius to be non-negative ($\rho \geq 0$) and the azimuth φ to lie in a specific interval spanning 360°, such as (−180°,+180°] or [0,360°).

Coordinate System Conversions

The cylindrical coordinate system is one of many three-dimensional coordinate systems. The following formula may be used to convert between them.

Cartesian Coordinates

For the conversion between cylindrical and Cartesian coordinate systems, it is convenient to assume that the reference plane of the former is the Cartesian x–y plane (with equation $z = 0$), and the cylindrical axis is the Cartesian z axis. Then the z coordinate is the same in both systems, and the correspondence between cylindrical (ρ, φ) and Cartesian (x, y) are the same as for polar coordinates, namely

$$x = \rho \cos \varphi$$

$$y = \rho \sin \varphi$$

in one direction, and

$$\rho = \sqrt{x^2 + y^2}$$

$$\varphi = \begin{cases} 0 & \text{if } x = 0 \text{ and } y = 0 \\ \arcsin(\frac{y}{\rho}) & \text{if } x \geq 0 \\ -\arcsin(\frac{y}{\rho}) + \pi & \text{if } x < 0 \end{cases}$$

in the other. The arcsin function is the inverse of the sine function, and is assumed to return an angle in the range [− π/2,+π/2] = [− 90°,+90°]. These formulas yield an azimuth φ in the range [− 90°,+270°].

Many modern programming languages provide a function that will compute the correct azimuth φ, in the range (−π, π], given x and y, without the need to perform a case analysis as above. For example, this function is called by atan2(y, x) in the C programming language, and atan(y, x) in Common Lisp.

Cylindrical Harmonics

The solutions to the Laplace equation in a system with cylindrical symmetry are called cylindrical harmonics.

Spherical Coordinate System

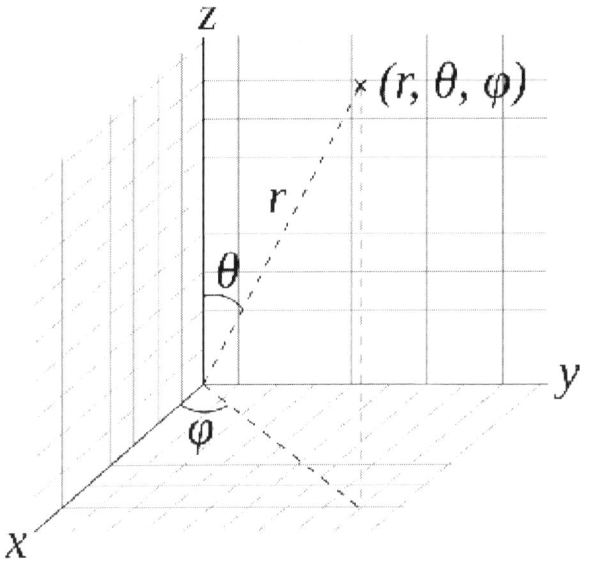

Spherical coordinates (r,θ,φ) as commonly used in *physics*: radial distance r, polar angle θ, and azimuthal angle φ. The symbol ρ is often used instead of r.

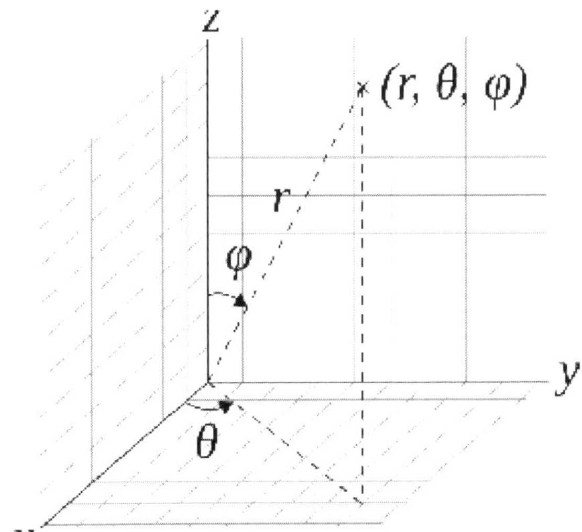

Spherical coordinates (r,θ,φ) as often used in *mathematics*: radial distance r, azimuthal angle θ, and polar angle φ. The meanings of θ and φ have been swapped compared to the physics convention.

In mathematics, a spherical coordinate system is a coordinate system for three-dimensional space where the position of a point is specified by three numbers: the radial distance of that point from a fixed origin, its polar angle measured from a fixed zenith direction, and the azimuth angle of its orthogonal projection on a reference plane that passes through the origin and is orthogonal to the zenith, measured from a fixed reference direction on that plane.

The radial distance is also called the radius or radial coordinate. The polar angle may be called colatitude, zenith angle, normal angle, or inclination angle. The use of symbols and the order of the coordinates differs between sources. In one system which is usual in physics gives the radial distance, polar angle, and azimuthal angle, whereas in another system used in many mathematics books gives the radial distance, azimuthal angle, and polar angle. In both systems is often used instead of . Other conventions are also used so great care needs to be taken to check which one is being used.

A number of different spherical coordinate systems are used outside mathematics which follow different conventions. In a geographical coordinate system positions are measured in latitude, longitude and height or altitude. There are a number of different celestial coordinate systems based on different fundamental planes and with different terms for the various coordinates. The spherical coordinate systems used in mathematics normally use radians rather than degrees and measure the azimuthal angle counter-clockwise rather than clockwise. The inclination angle is often replaced by the elevation angle measured from the reference plane. The concept of spherical coordinates can be extended to higher dimensional spaces and are then referred to as hyperspherical coordinates.

Definition

To define a spherical coordinate system, one must choose two orthogonal directions, the *zenith* and the *azimuth reference*, and an *origin* point in space. These choices determine a reference plane that contains the origin and is perpendicular to the zenith. The spherical coordinates of a point P are then defined as follows:

- The *radius* or *radial distance* is the Euclidean distance from the origin O to P.
- The *inclination* (or *polar angle*) is the angle between the zenith direction and the line segment OP.
- The *azimuth* (or *azimuthal angle*) is the signed angle measured from the azimuth reference direction to the orthogonal projection of the line segment OP on the reference plane.

The sign of the azimuth is determined by choosing what is a *positive* sense of turning about the zenith. This choice is arbitrary, and is part of the coordinate system's definition.

The *elevation* angle is 90 degrees ($\pi/2$ radians) minus the inclination angle. If the inclination is zero or 180 degrees (π radians), the azimuth is arbitrary. If the radius is zero, both azimuth and inclination are arbitrary. In linear algebra, the vector from the origin O to the point P is often called the *position vector* of P.

Conventions

Several different conventions exist for representing the three coordinates, and for the order in which they should be written. The use of (r, θ, φ) to denote, respectively, radial distance, inclination (or elevation), and azimuth, is common practice in physics, and is specified by ISO standard 31-11.

However, some authors (including mathematicians) use φ for inclination (or elevation) and θ for azimuth, which "provides a logical extension of the usual polar coordinates notation". Some authors may also list the azimuth before the inclination (or elevation), and/or use ρ instead of r for radial distance. Some combinations of these choices result in a left-handed coordinate system. The standard convention (r, θ, φ) conflicts with the usual notation for the two-dimensional polar coordinates, where θ is often used for the azimuth. It may also conflict with the notation used for three-dimensional cylindrical coordinates.

The angles are typically measured in degrees (°) or radians (rad), where $360° = 2\pi$ rad. Degrees are most common in geography, astronomy, and engineering, whereas radians are commonly used in mathematics and theoretical physics. The unit for radial distance is usually determined by the context.

When the system is used for physical three-space, it is customary to use positive sign for azimuth angles that are measured in the counter-clockwise sense from the reference direction on the reference plane, as seen from the zenith side of the plane. This convention is used, in particular, for geographical coordinates, where the "zenith" direction is north and positive azimuth (longitude) angles are measured eastwards from some prime meridian.

Unique Coordinates

Any spherical coordinate triplet (r, θ, φ) specifies a single point of three-dimensional space. On the other hand, every point has infinitely many equivalent spherical coordinates. One can add or subtract any number of full turns to either angular measure without changing the angles themselves, and therefore without changing the point. It is also convenient, in many contexts, to allow negative radial distances, with the convention that $(-r, \theta, \varphi)$ is equivalent to $(r, \theta + 180°, \varphi)$ for any r, θ, and φ. Moreover, $(r, -\theta, \varphi)$ is equivalent to $(r, \theta, \varphi + 180°)$.

If it is necessary to define a unique set of spherical coordinates for each point, one may restrict their ranges. A common choice is:

$$r > 0$$

$$0° \leq \theta \leq 180° \ (\pi \text{ rad})$$

$$0° \leq \varphi < 360° \ (2\pi \text{ rad})$$

However, the azimuth φ is often restricted to the interval $(-180°, +180°]$, or $(-\pi, +\pi]$ in radians, instead of $[0, 360°)$. This is the standard convention for geographic longitude.

The range $[0°, 180°]$ for inclination is equivalent to $[-90°, +90°]$ for elevation (latitude). Even with these restrictions, if θ is zero or 180° (elevation is 90° or -90°) then the azimuth angle is arbitrary; and if r is zero, both azimuth and inclination/elevation are arbitrary. To make the coordinates unique one can use the convention that in these cases the arbitrary coordinates are zero.

Plotting

To plot a point from its spherical coordinates (r, θ, φ), where θ is inclination, move r units from the origin in the zenith direction, rotate by θ about the origin towards the azimuth reference direction, and rotate by φ about the zenith in the proper direction.

Applications

The geographic coordinate system uses the azimuth and elevation of the spherical coordinate system to express locations on Earth, calling them respectively longitude and latitude. Just as the two-dimensional Cartesian coordinate system is useful on the plane, a two-dimensional spherical coordinate system is useful on the surface of a sphere. In this system, the sphere is taken as a unit sphere, so the radius is unity and can generally be ignored. This simplification can also be very useful when dealing with objects such as rotational matrices.

Spherical coordinates are useful in analysing systems that have some degree of symmetry about a point, such as volume integrals inside a sphere, the potential energy field surrounding a concentrated mass or charge, or global weather simulation in a planet's atmosphere.

A sphere that has the Cartesian equation $x^2 + y^2 + z^2 = c^2$ has the simple equation $r = c$ in spherical coordinates.

Two important partial differential equations that arise in many physical problems, Laplace's equation and the Helmholtz equation, allow a separation of variables in spherical coordinates. The angular portions of the solutions to such equations take the form of spherical harmonics. Another application is ergonomic design, where r is the arm length of a stationary person and the angles describe the direction of the arm as it reaches out. Three dimensional modelling of loudspeaker output patterns can be used to predict their performance. A number of polar plots are required, taken at a wide selection of frequencies, as the pattern changes greatly with frequency. Polar plots help to show that many loudspeakers tend towards omnidirectionality at lower frequencies.

The spherical coordinate system is also commonly used in 3D game development to rotate the camera around the player's position.

Coordinate System Conversions

As the spherical coordinate system is only one of many three-dimensional coordinate systems, there exist equations for converting coordinates between the spherical coordinate system and others.

Cartesian Coordinates

The spherical coordinates (*radius* r, *inclination* θ, *azimuth* φ) of a point can be obtained from its Cartesian coordinates (x, y, z) by the formulae

$$r = \sqrt{x^2 + y^2 + z^2}$$

$$\varphi = \arctan\left(\frac{y}{x}\right)$$

$$\theta = \arccos\left(\frac{z}{r}\right)$$

The inverse tangent denoted in $\varphi = \arctan(y/x)$ must be suitably defined, taking into account the correct quadrant of (x, y).

Alternatively, the conversion can be considered as two sequential rectangular to polar conversions: the first in the Cartesian x–y plane from (x, y) to (R,φ), where R is the projection of r onto the x–y plane, and the second in the Cartesian z-R plane from (z,R) to (r,θ). The correct quadrants for φ and θ are implied by the correctness of the planar rectangular to polar conversions.

These formulae assume that the two systems have the same origin, that the spherical reference plane is the Cartesian x–y plane, that θ is inclination from the z direction, and that the azimuth angles are measured from the Cartesian x axis (so that the y axis has $\varphi = +90°$). If θ measures elevation from the reference plane instead of inclination from the zenith the arccos above becomes an arcsin, and the cos θ and sin θ below become switched.

Conversely, the Cartesian coordinates may be retrieved from the spherical coordinates (*radius* r, *inclination* θ, *azimuth* φ), where $r \in [0, \infty), \varphi \in [0, 2\pi], \theta \in [0, \pi]$, by:

$$x = r\sin\theta\cos\phi$$

$$y = r\sin\theta\sin\phi$$

$$z = r\cos\theta$$

Geographic Coordinates

To a first approximation, the geographic coordinate system uses elevation angle (latitude) in degrees north of the equator plane, in the range $-90° \le \varphi \le 90°$, instead of inclination. Latitude is either geocentric latitude, measured at the Earth's centre and designated variously by ψ, q, φ', φ_c, φ_g or geodetic latitude, measured by the observer's local vertical, and commonly designated φ. The azimuth angle (longitude), commonly denoted by λ, is measured in degrees east or west from some conventional reference meridian (most commonly the IERS Reference Meridian), so its domain is $-180° \le \lambda \le 180°$. For positions on the Earth or other solid celestial body, the reference plane is usually taken to be the plane perpendicular to the axis of rotation.

The polar angle, which is 90° minus the latitude and ranges from 0 to 180°, is called colatitude in geography.

Instead of the radial distance, geographers commonly use altitude above some reference surface, which may be the sea level or "mean" surface level for planets without liquid oceans. The radial distance r can be computed from the altitude by adding the mean radius of the planet's reference surface, which is approximately 6,360 ± 11 km for Earth.

However, modern geographical coordinate systems are quite complex, and the positions implied by these simple formulae may be wrong by several kilometres. The precise standard meanings of latitude, longitude and altitude are currently defined by the World Geodetic System (WGS), and take into account the flattening of the Earth at the poles (about 21 km) and many other details.

In astronomy there are a series of spherical coordinate systems that measure the elevation angle from different fundamental planes. These reference planes are the observer's horizon, the celestial equator (defined by the Earth's rotation), the plane of the ecliptic (defined by Earth's orbit around the sun), and the galactic equator (defined by the rotation of the galaxy).

Kinematics

In spherical coordinates the position of a point is written,

$$\mathbf{r} = r\hat{\mathbf{r}},$$

its velocity is then,

$$\mathbf{v} = \dot{r}\hat{\mathbf{r}} + r\dot{\theta}\hat{\boldsymbol{\theta}} + r\dot{\varphi}\sin\theta\hat{\boldsymbol{\phi}}$$

and its acceleration is,

$$\mathbf{a} = \left(\ddot{r} - r\dot{\theta}^2 - r\dot{\varphi}^2\sin^2\theta\right)\hat{\mathbf{r}}$$

$$+\left(r\ddot{\theta} + 2\dot{r}\dot{\theta} - r\dot{\varphi}^2\sin\theta\cos\theta\right)\hat{\boldsymbol{\theta}}$$

$$+\left(r\ddot{\varphi}\sin\theta + 2\dot{r}\dot{\varphi}\sin\theta + 2r\dot{\theta}\dot{\varphi}\cos\theta\right)\hat{\boldsymbol{\phi}}$$

In the case of a constant φ this reduces to vector calculus in polar coordinates.

Homogeneous Coordinates

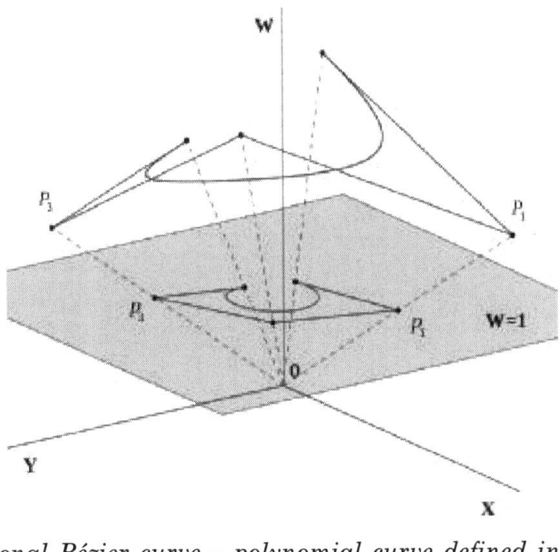

Figure: *Rational Bézier curve – polynomial curve defined in homogeneous coordinates (blue) and its projection on plane – rational curve (red)*

In mathematics, homogeneous coordinates, introduced by August Ferdinand Mφbius in his 1827 work *Der barycentrische Calcül*, are a system of coordinates used in projective geometry much as Cartesian coordinates are used in Euclidean geometry. They have the advantage that the coordinates of points, including points at infinity, can be represented using finite coordinates. Formulas involving homogeneous coordinates are often simpler and more symmetric than their Cartesian counterparts. Homogeneous coordinates have a range of applications, including computer graphics and 3D computer vision, where they allow affine transformations and, in general, projective transformations to be easily represented by a matrix.

If the homogeneous coordinates of a point are multiplied by a non-zero scalar then the resulting coordinates represent the same point. An additional condition must be added on the coordinates to ensure that only one set of coordinates corresponds to a given point, so the number of coordinates required is, in general, one more than the dimension of the projective space being considered. For example, two homogeneous coordinates are required to specify a point on the projective line and three homogeneous coordinates are required to specify a point on the projective plane.

Introduction

The projective plane can be thought of as the Euclidean plane with additional points, so called points at infinity, added. There is a point at infinity for each direction, informally defined as the limit of a point that moves in that direction away from a fixed point. Parallel lines in the Euclidean plane are said to intersect at a point at infinity corresponding to their common direction. A given point (x, y) on the Euclidean plane is identified with two ratios $(X/Z, Y/Z)$, so the point (x, y) corresponds to the triple $(X, Y, Z) = (xZ, yZ, Z)$ where $Z \neq 0$. Such a triple is a set of homogeneous coordinates for the point (x, y). Note that, since ratios are used, multiplying the three homogeneous coordinates by a common, non-zero factor does not change the point represented – unlike Cartesian coordinates, a single point can be represented by infinitely many homogeneous coordinates.

The equation of a line through the point (a, b) may be written $l(x - a) + m(y - b) = 0$ where l and m are not both 0. In parametric form this can be written $x = a + mt$, $y = b - lt$. Let $Z=1/t$, so the coordinates of a point on the line may be written $(a + m/Z, b - l/Z)=((aZ + m)/Z, (bZ - l)/Z)$. In homogeneous coordinates this becomes $(aZ + m, bZ - l, Z)$. In the limit as t approaches infinity, in other words as the point moves away from (a, b), Z becomes 0 and the homogeneous coordinates of the point become $(m, -l, 0)$. So $(m, -l, 0)$ are defined as homogeneous coordinates of the point at infinity corresponding to the direction of the line $l(x - a) + m(y - b) = 0$.

To summarize:

- Any point in the projective plane is represented by a triple (X, Y, Z), called the homogeneous coordinates of the point, where X, Y and Z are not all 0.
- The point represented by a given set of homogeneous coordinates is unchanged if the coordinates are multiplied by a common factor.
- Conversely, two sets of homogeneous coordinates represent the same point if and only if one is obtained from the other by multiplying by a common factor.
- When Z is not 0 the point represented is the point $(X/Z, Y/Z)$ in the Euclidean plane.
- When Z is 0 the point represented is a point at infinity.

Note that the triple (0, 0, 0) is omitted and does not represent any point. The origin is represented by (0, 0, 1).

Notation

Some authors use different notations for homogeneous coordinates which help distinguish them from Cartesian coordinates. The use of colons instead of commas, for example $(x:y:z)$ instead of (x, y, z), emphasises that the coordinates are to be considered ratios. Brackets, as in $[x, y, z]$ emphasise that multiple sets of coordinates are associated with a single point. Some authors use a combination of colons and brackets, as in $[x:y:z]$.

Homogeneity

Homogeneous coordinates are not uniquely determined by a point, so a function defined on the coordinates, say $f(x, y, z)$, does not determine a function defined on points as with Cartesian coordinates. But a condition $f(x, y, z) = 0$ defined on the coordinates, as might be used to describe a curve, determines a condition on points if the function is homogeneous. Specifically, suppose there is a k such that

$$f(\lambda x, \lambda y, \lambda z) = \lambda^k f(x, y, z).$$

If a set of coordinates represent the same point as (x, y, z) then it can be written $(\lambda x, \lambda y, \lambda z)$ for some non-zero value of λ. Then

$$f(x,y,z)=0 \Leftrightarrow f(\lambda x,\lambda y,\lambda z)=\lambda^k f(x,y,z)=0.$$

A polynomial $g(x, y)$ of degree k can be turned into a homogeneous polynomial by replacing x with x/z, y with y/z and multiplying by z^k, in other words by defining

$$f(x,y,z)=z^k g(x/z,y/z).$$

The resulting function f is a polynomial so it makes sense to extend its domain to triples where $z = 0$. The process can be reversed by setting $z = 1$, or

$$g(x,y)=f(x,y,1).$$

The equation $f(x, y, z) = 0$ can then be thought of as the homogeneous form of $g(x, y) = 0$ and it defines the same curve when restricted to the Euclidean plane. For example, the homogeneous form of the equation of the line $ax + by + c = 0$ is $ax + by + cz = 0$.

Other Dimensions

The discussions in the preceding sections apply analogously to projective spaces other than the plane. So the points on the projective line may be represented by pairs of coordinates (x, y), not both zero. In this case, the point at infinity is $(1, 0)$. Similarly the points in projective n-space are represented by $(n + 1)$-tuples.

Alternative Definition

Another definition of projective space can be given in terms of equivalence classes. For non-zero element of R^3, define $(x_1, y_1, z_1) \sim (x_2, y_2, z_2)$ to mean there is a non-zero λ so that $(x_1, y_1, z_1) = (\lambda x_2, \lambda y_2, \lambda z_2)$. Then $\sim$ is an equivalence relation and the projective plane can be defined as the equivalence classes of R^3 " $\{0\}$. If (x, y, z) is one of elements of the equivalence class p then these are taken to be homogeneous coordinates of p.

Lines in this space are defined to be sets of solutions of equations of the form $ax + by + cz = 0$ where not all of a, b and c are zero. The condition $ax + by + cz = 0$ depends only on the equivalence class of (x, y, z) so the equation defines a set of points in the projective line. The mapping $(x, y) \rightarrow (x, y, 1)$ defines an inclusion from the Euclidean plane to the projective plane and the complement of the image is the set of points with $z = 0$. This is the equation of a line according to the definition and the complement is called the *line at infinity*.

The equivalence classes, p, are the lines through the origin with the origin removed. The origin does not really play an essential part in the previous discussion so it can be added back in without changing the properties of the projective plane. This produces a variation on the definition, namely the projective plane is defined as the set of lines in R^3 that pass through the origin and the coordinates of a non-zero element (x, y, z) of a line are taken to be homogeneous coordinates of the line. These lines are now interpreted as points in the projective plane.

Again, this discussion applies analogously to other dimensions. So the projective space of dimension n can be defined as the set of lines through the origin in R^{n+1}.

Elements Other Than Points

The equation of a line in the projective plane may be given as $sx + ty + uz = 0$ where s, t and u are constants. Each triple (s, t, u) determines a line, the line determined is unchanged if it is multiplied by a nonzero scalar, and at least one of s, t and u must be non-zero. So the triple (s, t, u) may be taken to be homogeneous coordinates of a line in the projective plane, that is line coordinates as opposed to point coordinates.

If in $sx + ty + uz = 0$ the letters s, t and u are taken as variables and x, y and z are taken as constants then equation becomes an equation of a set of lines in the space of all lines in the plane.

Geometrically it represents the set of lines that pass though the point (x, y, z) and may be interpreted as the equation of the point in line-coordinates. In the same way, planes in 3-space may be given sets of four homogeneous coordinates, and so on for higher dimensions.

Duality (Projective Geometry)

A striking feature of projective planes is the "symmetry" of the roles played by points and lines in the definitions and theorems, and (plane) duality is the formalization of this metamathematical concept. There are two approaches to the subject of duality, one through language (the Principle of Duality) and the other a more functional approach.

These are completely equivalent and either treatment has as its starting point the axiomatic version of the geometries under consideration. In the functional approach there is a map between related geometries which is called a *duality*. In specific examples, such a map can be constructed in many ways. The concept of plane duality readily extends to space duality and beyond that to duality in any finite dimensional projective geometry.

Principle of Duality

If one defines a projective plane axiomatically as an incidence structure, in terms of a set P of points, a set L of lines, and an incidence relation I that determines which points lie on which lines, then one may define a plane dual structure.

Interchange the role of "points" and "lines" in

$$C = (P, L, I)$$

to obtain the dual structure

$$C^* = (L, P, I^*),$$

where I^* is the inverse relation of I. C* is also a projective plane, called the dual plane of C.

If C and C* are isomorphic, then C is called *self-dual*. The projective planes PG(2, K) for any division ring K are self-dual. However, there are non-Desarguesian planes which are not self-dual, such as the Hall planes and some that are, such as the Hughes planes.

In a projective plane a statement involving points, lines and incidence between them that is obtained from another such statement by interchanging the words "point" and "line" and making whatever grammatical adjustments that are necessary, is called the plane dual statement of the first.

The plane dual statement of "Two points are on a unique line." is "Two lines meet at a unique point." Forming the plane dual of a statement is known as *dualising* the statement.

If a statement is true in a projective plane C, then the plane dual of that statement must be true in the dual plane C*. This follows since dualising each statement in the proof "in C– gives a statement of the proof "in C*."

The *Principle of Plane Duality* says that dualising any theorem in a self-dual projective plane C produces another theorem valid in C.

The above concepts can be generalised to talk about space duality, where the terms "points" and "planes" are interchanged (and lines remain lines). This leads to the *Principle of Space Duality*. Further generalisation is possible.

These principles provide a good reason for preferring to use a "symmetric" term for the incidence relation. Thus instead of saying "a point lies on a line" one should say "a point is incident with a line" since dualising the latter only involves interchanging point and line ("a line is incident with a point").

Traditionally in projective geometry, the set of points on a line are considered to include the relation of projective harmonic conjugates. In this tradition the points on a line form a projective range, a concept dual to a pencil of lines on a point.

Dual Theorems

As the real projective plane, PG(2, R), is self-dual there are a number of pairs of well known results that are duals of each other. Some of these are:

- Desargues' theorem ⇔ Converse of Desargues' theorem
- Pascal's theorem ⇔ Brianchon's theorem
- Menelaus' theorem ⇔ Ceva's theorem

Duality as a Mapping

A (plane) duality is a map from a projective plane C = (P, L, I) to its dual plane C* = (L, P, I*) which preserves incidence. That is, a (plane) duality σ will map points to lines and lines to points (P^σ = L and L^σ = P) in such a way that if a point Q is on a line m (denoted by Q I m) then Q^σ I* m^σ Ô! m^σ I Q^σ. A (plane) duality which is an isomorphism is called a correlation.[1] The existence of a correlation means that the projective plane C is self-dual. In the special case that the projective plane is of the PG(2, *K*) type, with *K* a division ring, a duality is called a reciprocity. These planes are always self-dual. By the Fundamental theorem of projective geometry a reciprocity is the composition of an automorphic function of *K* and a homography. If the automorphism involved is the identity, then the reciprocity is called a projective correlation. A correlation of order two (an involution) is called a polarity. If a correlation φ is not a polarity then φ^2 is a nontrivial collineation. This duality mapping concept can also be extended to higher dimensional spaces so the modifier "(plane)" can be dropped in those situations.

Higher Dimensional Duality

Duality in the projective plane is a special case of duality for projective spaces, transformations of PG(n, *K*) (also denoted by KP^n) with *K* a field, that interchange objects of dimension *r* with objects of dimension $n - 1 - r$ (= codimension r + 1). That is, in a projective space of dimension *n*, the points (dimension 0) are made to correspond with hyperplanes (codimension 1), the lines joining two points (dimension 1) are made to correspond with the intersection of two hyperplanes (codimension 2), and so on.

The points of PG(n,K) can be taken to be the nonzero vectors in the $(n + 1)$-dimensional vector space over K, where we identify two vectors which differ by a scalar factor. Another way to put it is that the points of n-dimensional projective space are the lines through the origin in K^{n+1}, which are 1-dimensional vector subspaces. Also the n- vector dimensional subspaces of K^{n+1} represent the $(n - 1)$- geometric dimensional hyperplanes of projective n-space over K.

A nonzero vector u = $(u_0, u_1,..., u_n)$ in K^{n+1} also determines an (n - 1) - geometric dimensional subspace (hyperplane) H_u, by

$$H_u = \{(x_0,x_1,...,x_n) : u_0x_0 + ... + u_nx_n = 0\}.$$

When a vector u is used to define a hyperplane in this way it shall be denoted by u_H, while if it is designating a point we will use u_P. In terms of the usual dot product, $H_u = \{x_P : u_H \cdot x_P = 0\}$. Since K is a field, the dot product is symmetrical, meaning $u_H \cdot x_P = u_0x_0 + u_1x_1 + ... + u_nx_n = x_0u_0 + x_1u_1 + ... + x_nu_n = x_H \cdot u_P$. A reciprocity can be given by $u_P \leftrightarrow H_u$ between points and hyperplanes. This extends to a reciprocity between the line generated by two points and the intersection of two such hyperplanes, and so forth.

In the projective plane, PG(2,K), with K a field we have the reciprocity given by: points in homogeneous coordinates $(a,b,c) \leftrightarrow$ lines with equations $ax + by + cz = 0$. In a corresponding projective space, PG(3,K), a reciprocity is given by: points in homogeneous coordinates $(a,b,c,d) \leftrightarrow$ planes with equations $ax + by + cz + dw = 0$. This reciprocity would also map a line determined by two points (a_1,b_1,c_1,d_1) and (a_2,b_2,c_2,d_2) to the line which is the intersection of the two planes with equations $a_1x + b_1y + c_1z + d_1w = 0$ and $a_2x + b_2y + c_2z + d_2w = 0$.

Three Dimensions

In a polarity of real projective 3-space, PG(3,R), points correspond to planes, and lines correspond to lines. By restriction the duality of polyhedra in solid geometry is obtained, where points are dual to faces, and sides are dual to sides, so that the icosahedron is dual to the dodecahedron, and the cube is dual to the octahedron.

Geometric Construction of a Reciprocity

The reciprocity of PG(2,K), with K a field, given by homogeneous coordinates can also be described geometrically. This uses the model of the real projective plane which is a "unit sphere with antipodes identifie ≤, or equivalently, the model of lines and planes through the origin of the vector space K^3. Associate a line through the origin with the unique plane through the origin which is perpendicular (orthogonal) to the line. When, in the model, these lines are considered to be the points and the planes the lines of the projective plane PG(2,*K*), this association becomes a reciprocity (actually a polarity) of the projective plane. The sphere model is obtained by intersecting the lines and planes through the origin with a unit sphere centreed at the origin. The lines meet the sphere in antipodal points which must then be identified to obtain a point of the projective plane, and the planes meet the sphere in great circles which are thus the lines of the projective plane.

That this association "preserves" incidence is most easily seen from the lines and planes model. A point incident with a line in the projective plane corresponds to a line lying in a plane in the model. Applying the association, the plane becomes a line through the origin perpendicular to the plane it is associated with. This image line is perpendicular to every line of the plane which passes through the origin, in particular the original line (point of the projective plane). All lines that are perpendicular to the original line at the origin lie in the unique plane which is orthogonal to the original line, that is, the image plane under the association. Thus, the image line lies in the image plane and the association preserves incidence.

Poles and Polars

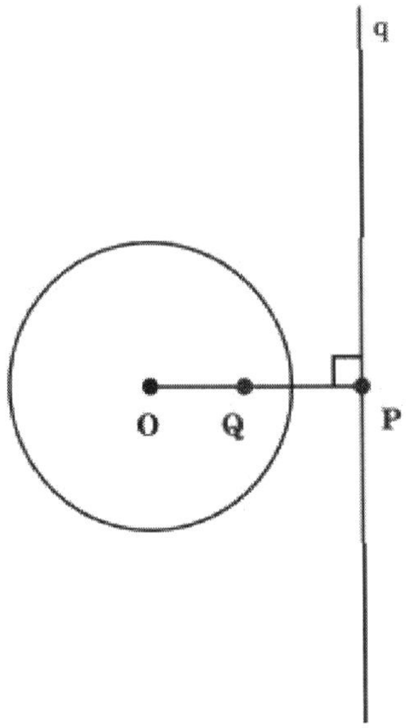

Figure: *Pole and polar with respect to circle O. P = Q', q is polar of Q, Q is pole of q.*

In the Euclidean plane, fix a circle C with centre O and radius r. For each point P other than O define an image point P *so that* OP • OP = r^2.

The mapping defined by $P \rightarrow P$ *is called inversion with respect to circle* C. *The line through* P which is perpendicular to the line OP is called the polar of the point P with respect to circle C. Let m be a line not passing through O. Drop a perpendicular from O to m, meeting m at the point Q (this is the point of m that is closest to O).

The image of Q under inversion with respect to C is called the pole of m. If a point P (different from O) is on a line m (not passing through O) then the pole of m lies on the polar of P and viceversa. The incidence preserving process, in which points and lines are transformed into their polars and poles with respect to C is called reciprocation.

In order to turn this process into a reciprocity, the Euclidean plane (which is not a projective plane) needs to be expanded to the extended euclidean plane by adding a line at infinity and points at infinity which lie on this line.

In this expanded plane, we define the polar of the point O to be the line at infinity (and O is the pole of the line at infinity), and the poles of the lines through O are the points of infinity where, if a line has slope s ($\neq 0$) its pole is the infinite point associated to the parallel class of lines with slope $-1/s$. The pole of the x-axis is the point of infinity of the vertical lines and the pole of the y-axis is the point of infinity of the horizontal lines.

The construction of a reciprocity based on inversion in a circle given above can be generalised by using inversion in a conic section (in the extended real plane). The reciprocities constructed in this manner are projective correlations of order two, that is, polarities.

Mapping the Sphere onto the Plane

The unit sphere modulo “1 model of the projective plane is isomorphic (w.r.t. incidence properties) to the planar model: the affine plane extended with a projective line at infinity.

To map a point on the sphere to a point on the plane, let the plane be tangent to the sphere at some point which shall be the origin of the plane’s coordinate system (2-D origin). Then construct a line passing through the centre of the sphere (3-D origin) and the point on the sphere. This line intersects the plane at a point which is the projection of the point on the sphere onto the plane (or vice versa).

This projection can be used to define a one-to-one onto mapping

$$f:[0,\pi/2]\times[0,2\pi]\rightarrow \mathbb{R}P^2.$$

If points in $\mathbb{R}P^2$ are expressed in homogeneous coordinates, then

$$f:(\theta,\phi)\mapsto[\cos\phi:\sin\phi:\cot\theta],$$

$$f^{-1}:[x:y:z]\mapsto\left(\arctan\sqrt{\left(\frac{x}{z}\right)^2+\left(\frac{y}{z}\right)^2},\arctan_2(y,x)\right).$$

Also, lines in the planar model are projections of great circles of the sphere. This is so because through any line in the plane pass an infinitude of different planes: one of these planes passes through the 3-D origin, but a plane passing through the 3-D origin intersects the sphere along a great circle. As we have seen, any great circle in the unit sphere has a projective point perpendicular to it, which can be defined as its dual. But this point is a pair of antipodal points on the unit sphere, through both of which passes a unique 3-D line, and this line extended past the unit sphere intersects the tangent plane at a point, which means that there is a geometric way to associate a unique point on the plane to every line on the plane, such that the point is the dual of the line.

Plücker Coordinates

In geometry, Plücker coordinates, introduced by Julius Plücker in the 19th century, are a way to assign six homogenous coordinates to each line in projective 3-space, P^3. Because they satisfy a quadratic constraint, they establish a one-to-one correspondence between the 4-dimensional space of lines in P^3 and points on a quadric in P^5 (projective 5-space). A predecessor and special case of Grassmann coordinates (which describe k-dimensional linear subspaces, or *flats*, in an n-dimensional Euclidean space), Plücker coordinates arise naturally in geometric algebra. They have proved useful for computer graphics, and also can be extended to coordinates for the screws and wrenches in the theory of kinematics used for robot control.

Geometric Intuition

A line L in 3-dimensional Euclidean space is determined by two distinct points that it contains, or by two distinct planes that contain it. Consider the first case, with points $x = (x_1,x_2,x_3)$ and $y = (y_1,y_2,y_3)$. The vector displacement from x to y is nonzero because the points are distinct, and represents the *direction* of the line. That is, every displacement between points on L is a scalar multiple of $d = y–x$. If a physical particle of unit mass were to move from x to y, it would have a moment about the origin. The geometric equivalent is a vector whose direction is perpendicular to the plane containing L and the origin, and whose length equals twice the area of the triangle formed by the displacement and the origin. Treating the points as displacements from the origin, the moment is $m = x \times y$, where "×" denotes the vector cross product. The area of the triangle is proportional to the length of the segment between x and y, considered as the base of the triangle; it is not changed by sliding the base along the line, parallel to itself. By definition the moment vector is perpendicular to every displacement along the line, so $d \cdot m = 0$, where "•" denotes the vector dot product.

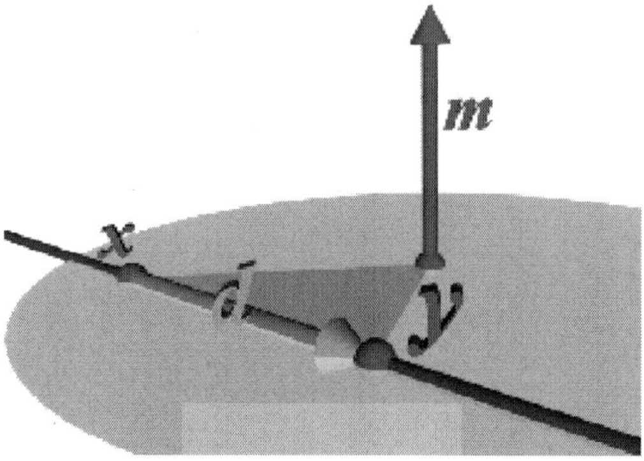

***Figure:** Displacement and moment of two points on line*

Although neither d nor m alone is sufficient to determine L, together the pair does so uniquely, up to a common (nonzero) scalar multiple which depends on the distance between x and y. That is, the coordinates

$$(d{:}m) = (d_1{:}d_2{:}d_3{:}m_1{:}m_2{:}m_3)$$

may be considered homogeneous coordinates for L, in the sense that all pairs $(\lambda d{:}\lambda m)$, for $\lambda \neq 0$, can be produced by points on L and only L, and any such pair determines a unique line so long as d is not zero and $d \cdot m = 0$. Furthermore, this approach extends to include points, lines, and a plane "at infinity", in the sense of projective geometry.

Example: Let $x = (2,3,7)$ and $y = (2,1,0)$. Then $(d{:}m) = (0{:}-2{:}-7{:}-7{:}14{:}-4)$.

Alternatively, let the equations for points x of two distinct planes containing L be

$$0 = a + a \cdot x$$

$$0 = b + b \cdot x\ .$$

Then their respective planes are perpendicular to vectors a and b, and the direction of L must be perpendicular to both. Hence we may set $d = a \times b$, which is nonzero because a and b are neither zero nor parallel (the planes being distinct and intersecting). If point x satisfies both plane equations, then it also satisfies the linear combination

$$0 = a\ (b + b \cdot x) - b\ (a + a \cdot x)$$
$$= (a\ b - b\ a) \cdot x\ .$$

That is, $m = a\ b - b\ a$ is a vector perpendicular to displacements to points on L from the origin; it is, in fact, a moment consistent with the d previously defined from a and b.

Example: Let $a_0 = 2$, $a = (-1,0,0)$ and $b_0 = -7$, $b = (0,7,-2)$. Then $(d{:}m) = (0{:}-2{:}-7{:}-7{:}14{:}-4)$.

Although the usual algebraic definition tends to obscure the relationship, $(d{:}m)$ are the Plücker coordinates of L.

Algebraic Definition

In a 3-dimensional projective space, P^3, let L be a line containing distinct points x and y with homogeneous coordinates $(x_0{:}x_1{:}x_2{:}x_3)$ and $(y_0{:}y_1{:}y_2{:}y_3)$, respectively. Let M be the 4×2 matrix with these coordinates as columns.

$$M = \begin{bmatrix} x_0 & y_0 \\ x_1 & y_1 \\ x_2 & y_2 \\ x_3 & y_3 \end{bmatrix}$$

Because x and y are distinct points, the columns of M are linearly independent; M has rank 2. Let M' be a second matrix, with columns x' and y'' a different pair of distinct points on L. Then the columns of M' are linear combinations of the columns of M; so for some 2×2 nonsingular matrix Λ,

$$M' = M\Lambda.$$

In particular, rows i and j of $M2$ and M are related by

$$\begin{bmatrix} x_i' & y_i' \\ x_j' & y_j' \end{bmatrix} = \begin{bmatrix} x_i & y_i \\ x_j & y_j \end{bmatrix} \begin{bmatrix} \lambda_{00} & \lambda_{01} \\ \lambda_{10} & \lambda_{11} \end{bmatrix}.$$

Therefore, the determinant of the left side 2 × 2 matrix equals the product of the determinants of the right side 2 × 2 matrices, the latter of which is a fixed scalar, det Λ.

Primary Coordinates

With this motivation, we define Plücker coordinate p_{ij} as the determinant of rows i and j of M,

$$\begin{aligned} p_{ij} &= \begin{vmatrix} x_i & y_i \\ x_j & y_j \end{vmatrix} \\ &= x_i y_j - x_j y_i . \end{aligned}$$

This implies $p_{ii} = 0$ and $p_{ij} = -p_{ji}$, reducing the possibilities to only six (4 choose 2) independent quantities. As we have seen, the sixtuple

$$(p_{01} : p_{02} : p_{03} : p_{23} : p_{31} : p_{12})$$

is uniquely determined by L, up to a common nonzero scale factor. Furthermore, all six components cannot be zero, because if they were, all 2 × 2 subdeterminants in M would be zero and the rank of M at most one, contradicting the assumption that x and y are distinct. Thus the Plücker coordinates of L, as suggested by the colons, may be considered homogeneous coordinates of a point in a 5-dimensional projective space.

Plücker Map

Denote the set of all lines (linear images of P^1) in P^3 by $G_{1,3}$. We thus have a map:

$$\begin{aligned} \alpha : G_{1,3} &\to \mathbf{P}^5 \\ L &\mapsto L^\alpha , \end{aligned}$$

where

$$L^\alpha = (p_{01} : p_{02} : p_{03} : p_{23} : p_{31} : p_{12}).$$

Dual Coordinates

Alternatively, let L be a line contained in distinct planes a and b with homogeneous coefficients $(a^0{:}a^1{:}a^2{:}a^3)$ and $(b^0{:}b^1{:}b^2{:}b^3)$, respectively. (The first plane equation is $0 = -_k a^k x_k$, for example.) Let N be the 2 × 4 matrix with these coordinates as rows.

$$N = \begin{bmatrix} a^0 & a^1 & a^2 & a^3 \\ b^0 & b^1 & b^2 & b^3 \end{bmatrix}$$

We define dual Plücker coordinate p^{ij} as the determinant of columns i and j of N,

$$p^{ij} = \begin{vmatrix} a^i & a^j \\ b^i & b^j \end{vmatrix}$$

$$= a^i b^j - a^j b^i .$$

Dual coordinates are convenient in some computations, and we can show that they are equivalent to primary coordinates. Specifically, let (i,j,k,l) be an even permutation of (0,1,2,3); then

$$p_{ij} = p^{kl}.$$

Geometry

To relate back to the geometric intuition, take $x_0 = 0$ as the plane at infinity; thus the coordinates of points *not* at infinity can be normalized so that $x_0 = 1$. Then M becomes

$$M = \begin{bmatrix} 1 & 1 \\ x_1 & y_1 \\ x_2 & y_2 \\ x_3 & y_3 \end{bmatrix},$$

and setting $x = (x_1,x_2,x_3)$ and $y = (y_1,y_2,y_3)$, we have $d = (p_{01},p_{02},p_{03})$ and $m = (p_{23},p_{31},p_{12})$. Dually, we have $d = (p^{23},p^{31},p^{12})$ and $m = (p^{01},p^{02},p^{03})$.

Point Equations

Letting $(x_0{:}x_1{:}x_2{:}x_3)$ be the point coordinates, four possible points on a line each have coordinates $x_i = p_{ij}$, for $j = 0\ldots3$. Some of these possible points may be inadmissible because all coordinates are zero, but since at least one Plücker coordinate is nonzero, at least two distinct points are guaranteed.

Bijectivity

If $(q_{01}{:}q_{02}{:}q_{03}{:}q_{23}{:}q_{31}{:}q_{12})$ are the homogeneous coordinates of a point in P^5, without loss of generality assume that q_{01} is nonzero. Then the matrix

$$M = \begin{bmatrix} q_{01} & 0 \\ 0 & q_{01} \\ -q_{12} & q_{02} \\ q_{31} & q_{03} \end{bmatrix}$$

has rank 2, and so its columns are distinct points defining a line L. When the P^5 coordinates, q_{ij}, satisfy the quadratic Plücker relation, they are the Plücker coordinates of L. To see this, first normalise q_{01} to 1. Then we immediately have that for the Plücker coordinates computed from M, $p_{ij} = q_{ij}$, except for

$$p_{23} = -q_{03}q_{12} - q_{02}q_{31}.$$

But if the q_{ij} satisfy the Plücker relation $q_{23}+q_{02}q_{31}+q_{03}q_{12}=0$, then $p_{23}=q_{23}$, completing the set of identities.

Consequently, α is a surjection onto the algebraic variety consisting of the set of zeros of the quadratic polynomial

$$p_{01}p_{23}+p_{02}p_{31}+p_{03}p_{12}\,.$$

And since α is also an injection, the lines in P^3 are thus in bijective correspondence with the points of this quadric in P^5, called the Plücker quadric or Klein quadric.

Uses

Plücker coordinates allow concise solutions to problems of line geometry in 3-dimensional space, especially those involving incidence.

Line-line Join

In the event that two lines are coplanar but not parallel, their common plane has equation

$$0 = (m \cdot d2\ )x_0 + (d \times d2\ ) \cdot x\ ,$$

where $x = (x_1, x_2, x_3)$.

The slightest perturbation will destroy the existence of a common plane, and near-parallelism of the lines will cause numeric difficulties in finding such a plane even if it does exist.

Line-line Meet

Dually, two coplanar lines, neither of which contains the origin, have common point

$$(x_0 : x) = (\mathrm{d} \cdot \mathrm{m2}\ : \mathrm{m} \times \mathrm{m2}\ )\ .$$

To handle lines not meeting this restriction.

Plane-line Meet

Given a plane with equation

$$0 = a^0 x_0 + a^1 x_1 + a^2 x_2 + a^3 x_3,$$

or more concisely $0 = a^0 x_0 + a \cdot x$; and given a line not in it with Plücker coordinates $(d{:}m)$, then their point of intersection is

$$(x_0 : x) = (a \cdot d : a \times m - a_0 d)\ .$$

The point coordinates, $(x_0{:}x_1{:}x_2{:}x_3)$, can also be expressed in terms of Plücker coordinates as

$$x_i = \sum_{j \neq i} a^j p_{ij}, \qquad i = 0 \ldots 3.$$

Point-line Join

Dually, given a point $(y_0{:}y)$ and a line not containing it, their common plane has equation

$$0 = (y \cdot m)\ x_0 + (y \times \leq y_0 m) \cdot x\ .$$

The plane coordinates, $(a^0{:}a^1{:}a^2{:}a^3)$, can also be expressed in terms of dual Plücker coordinates as

$$a^i = \sum_{j \neq i} y_j p^{ij}, \qquad i = 0 \ldots 3.$$

Line Families

Because the Klein quadric is in P^5, it contains linear subspaces of dimensions one and two (but no higher). These correspond to one- and two-parametre families of lines in P^3. For example, suppose L and L' are distinct lines in P^3 determined by points x, y and x′, y′, respectively. Linear combinations of their determining points give linear combinations of their Plücker coordinates, generating a one-parametre family of lines containing L and L'. This corresponds to a one-dimensional linear subspace belonging to the Klein quadric.

Lines in Plane

If three distinct and non-parallel lines are coplanar; their linear combinations generate a two-parametre family of lines, all the lines in the plane. This corresponds to a two-dimensional linear subspace belonging to the Klein quadric.

Lines Through Point

If three distinct and non-coplanar lines intersect in a point, their linear combinations generate a two-parametre family of lines, all the lines through the point. This also corresponds to a two-dimensional linear subspace belonging to the Klein quadric.

Ruled Surface

A ruled surface is a family of lines that is not necessarily linear. It corresponds to a curve on the Klein quadric. For example, a hyperboloid of one sheet is a quadric surface in P^3 ruled by two different families of lines, one line of each passing through each point of the surface; each family corresponds under the Plücker map to a conic section within the Klein quadric in P^5.

Line Geometry

During the nineteenth century, *line geometry* was studied intensively. In terms of the bijection given above, this is a description of the intrinsic geometry of the Klein quadric.

Ray Tracing

Line geometry is extensively used in ray tracing application where the geometry and intersections of rays need to be calculated in 3D. An implementation is described in Introduction to Pluecker Coordinates written for the Ray Tracing forum by Thouis Jones.

Application to Bézout's Theorem

Bézout's theorem predicts that the number of points of intersection of two curves is equal to the product of their degrees (assuming an algebraically complete field and with certain conventions followed for counting intersection multiplicities). Bézout's theorem predicts there is one point of intersection of two lines and in general this is true, but when the lines are parallel the point of intersection is infinite. Homogeneous coordinates can be used to locate the point of intersection in this case. Similarly, Bézout's theorem predicts that a line will intersect a conic at two points, but in some cases one or both of the points is infinite and homogeneous coordinates must be used to locate them. For example, $y = x^2$ and $x = 0$ have only one point of intersection in the finite plane. To find the other point of intersection, convert the equations into homogeneous form, $yz = x^2$ and $x = 0$. This produces $x = yz = 0$ and, assuming not all of x, y and z are 0, the solutions are $x = y = 0$, $z \neq 0$ and $x = z = 0$, $y \neq 0$. This first solution is the point (0, 0) in Cartesian coordinates, the finite point of intersection. The second solutions gives the homogeneous coordinates (0, 1, 0) which corresponds to the direction of the y-axis. For the equations $xy = 1$ and $x = 0$ there are no finite points of intersection. Converting the equations into homogeneous form gives $xy = z^2$ and $x = 0$. Solving produces the equation $z^2 = 0$ which has a double root at $z = 0$. From the original equation, $x = 0$, so $y \neq 0$ since at least one coordinate must be non-zero. Therefore (0, 1, 0) is the point of intersection counted with multiplicity 2 in agreement with the theorem.Circular points at infinity

In projective geometry, the circular points at infinity in the complex projective plane (also called cyclic points or isotropic points) are

$$(1: i: 0) \text{ and } (1: -i: 0).$$

Here the coordinates are homogeneous coordinates $(x: y: z)$; so that the line at infinity is defined by $z = 0$. These points at infinity are called circular points at infinity because they lie on the complexification of every real circle. In other words, both points satisfy the homogeneous equations of the type

$$Ax^2 + Ay^2 + 2B_1xz + 2B_2yz - Cz^2 = 0.$$

The case where the coefficients are all real gives the equation of a general circle (of the real projective plane). In general, an algebraic curve that passes through these two points is called circular. The circular points at infinity are the points at infinity of the isotropic lines. The circular points are invariant under translation and rotation.

Change of Coordinate Systems

Just as the selection of axes in the Cartesian coordinate is somewhat arbitrary, the selection of a single system of homogeneous coordinates out of all possible systems is somewhat arbitrary. Therefore it is useful to know how the different systems are related to each other.

Let (x, y, z) be the homogeneous coordinates of a point in the projective plane and for a fixed matrix

$$A = \begin{pmatrix} a & b & c \\ d & e & f \\ g & h & i \end{pmatrix},$$

with $\det(A) \neq 0$, define a new set of coordinates (X, Y, Z) by the equation

$$\begin{pmatrix} X \\ Y \\ Z \end{pmatrix} = A \begin{pmatrix} x \\ y \\ z \end{pmatrix}.$$

Multiplication of (x, y, z) by a scalar results in the multiplication of (X, Y, Z) by the same scalar, and X, Y and Z cannot be all 0 unless x, y and z are all zero since A is nonsingular. So (X, Y, Z) are a new system of homogeneous coordinates for points in the projective plane. If z is fixed at 1 then

$$X = ax + by + c, Y = dx + ey + f, Z = gx + hy + i$$

are proportional to the signed distances from the point to the lines

$$ax + by + c = 0, dx + ey + f = 0, gx + hy + i = 0.$$

(The *signed distance* is the distance multiplied a sign 1 or -1 depending on which side of the line the point lies.) Note that for $a = b = 0$ the value of X is simply a constant, and similarly for Y and Z.

The three lines,

$$ax + by + cz = 0, dx + ey + fz = 0, gx + hy + iz = 0$$

in homogeneous coordinates, or

$$X = 0, Y = 0, Z = 0$$

in the (*X*, *Y*, *Z*) system, form a triangle called the *triangle of reference* for the system.

Manifold

In mathematics (specifically in geometry and topology), a manifold is a geometrical object that, near each point, resembles Euclidean space of a fixed dimension, called the dimension of the manifold. In mathematical terms, a manifold of dimension n is a topological space such that each point has a neighbourhood that is homeomorphic to the Euclidean space of dimension n. Lines and circles (but not figure eights) are one-dimensional manifolds (1-manifolds). The plane, the sphere and the torus are examples of 2-manifolds.

Although near each point, a manifold resembles Euclidean space, globally a manifold might not. For example, the surface of the sphere is not a Euclidean space, but in a region it can be charted by means of geographic maps: map projections of the region into the Euclidean plane. When a region appears in two neighbouring maps (in the context of manifolds they are called *charts*), the two representations do not coincide exactly and a transformation is needed to pass from one to the other, called a *transition map*.

The concept of a manifold is central to many parts of geometry and modern mathematical physics because it allows more complicated structures to be described and understood in terms of the relatively well-understood properties of Euclidean space. Manifolds naturally arise as solution sets of systems of equations and as graphs of functions. Manifolds may have additional features. One important class of manifolds is the class of differentiable manifolds. This differentiable structure allows calculus to be done on manifolds. A Riemannian metric on a manifold allows distances and angles to be measured. Symplectic manifolds serve as the phase spaces in the Hamiltonian formalism of classical mechanics, while four-dimensional Lorentzian manifolds model space-time in general relativity.

Motivational Examples

Circle

After a line, the circle is the simplest example of a topological manifold. Topology ignores bending, so a small piece of a circle is treated exactly the same as a small piece of a line. Consider, for instance, the top half of the unit circle, $x^2 + y^2 = 1$, where the y-coordinate is positive (indicated by the yellow arc in *Figure 1*). Any point of this semicircle can be uniquely described by its x-coordinate. So, projection onto the first coordinate is a continuous, and invertible, mapping from the upper semicircle to the open interval (– 1,1):

$$\chi_{\text{top}}(x,y) = x.$$

Such functions along with the open regions they map are called *charts*. Similarly, there are charts for the bottom (red), left (blue), and right (green) parts of the circle.

Together, these parts cover the whole circle and the four charts form an atlas for the circle.

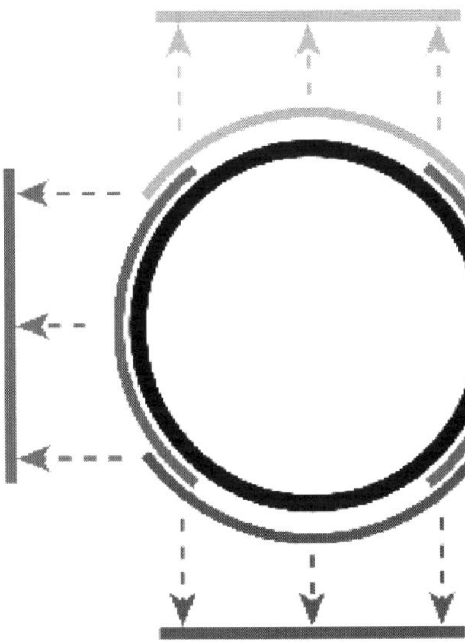

Figure: *The four charts each map part of the circle to an open interval, and together cover the whole circle.*

The top and right charts overlap: their intersection lies in the quarter of the circle where both the x- and the y-coordinates are positive. The two charts χ_{top} and χ_{right} each map this part into the interval (0, 1). Thus a function T from (0, 1) to itself can be constructed, which first uses the inverse of the top chart to reach the circle and then follows the right chart back to the interval. Let a be any number in (0, 1), then:

$$\begin{aligned} T(a) &= \chi_{\text{right}}\left(\chi_{\text{top}}^{-1}\left[a\right]\right) \\ &= \chi_{\text{right}}\left(a,\sqrt{1-a^2}\right) \\ &= \sqrt{1-a^2} \end{aligned}$$

Such a function is called a *transition map.*

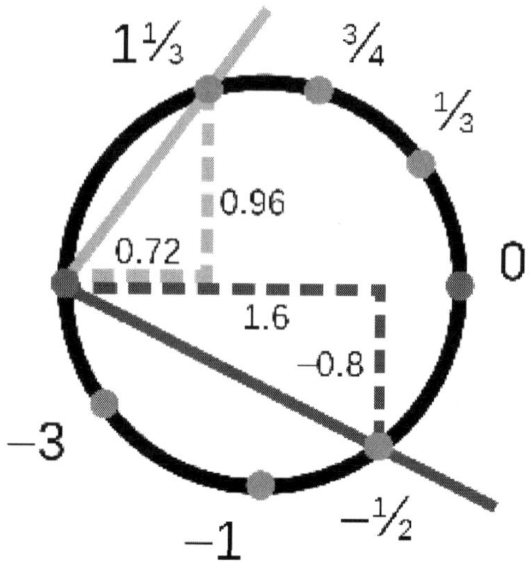

Figure: *A circle manifold chart based on slope, covering all but one point of the circle.*

The top, bottom, left, and right charts show that the circle is a manifold, but they do not form the only possible atlas. Charts need not be geometric projections, and the number of charts is a matter of some choice. Consider the charts

$$\chi_{\text{minus}}(x, y) = s = \frac{y}{1+x}$$

and

$$\chi_{\text{plus}}(x, y) = t = \frac{y}{1-x}$$

Here s is the slope of the line through the point at coordinates (x, y) and the fixed pivot point $(-1, 0)$; t is the mirror image, with pivot point $(+1, 0)$. The inverse mapping from s to (x, y) is given by

$$x = \frac{1-s^2}{1+s^2}$$

$$y = \frac{2s}{1+s^2}$$

It can easily be confirmed that $x^2 + y^2 = 1$ for all values of the slope s. These two charts provide a second atlas for the circle, with

$$t = \frac{1}{s}$$

Each chart omits a single point, either (–1, 0) for s or (+1, 0) for t, so neither chart alone is sufficient to cover the whole circle. It can be proved that it is not possible to cover the full circle with a single chart. For example, although it is possible to construct a circle from a single line interval by overlapping and "gluing" the ends, this does not produce a chart; a portion of the circle will be mapped to both ends at once, losing invertibility.

Other Curves

Manifolds need not be connected (all in "one piece"); an example is a pair of separate circles. In this example we see that a manifold need not have any well-defined notion of distance, for there is no way to define the distance between points that don't lie in the same piece. Manifolds need not be closed; thus a line segment without its end points is a manifold. And they are never countable, unless the dimension of the manifold is 0. Putting these freedoms together, other examples of manifolds are a parabola, a hyperbola (two open, infinite pieces) and the locus of points on a cubic curve $y^2 = x^3 - x$ (a closed loop piece and an open, infinite piece). However, we exclude examples like two touching circles that share a point to form a figure-8; at the shared point we cannot create a satisfactory chart. Even with the bending allowed by topology, the vicinity of the shared point looks like a "+", not a line (a + is not homeomorphic to a closed interval (line segment) since deleting the centre point from the + gives a space with four components (i.e. pieces) whereas deleting a point from a closed interval gives a space with at most two pieces; topological operations always preserve the number of pieces).

Enriched Circle

Viewed using calculus, the circle transition function T is simply a function between open intervals, which gives a meaning to the statement that T is differentiable. The transition map T, and all the others, are differentiable on (0, 1); therefore, with this atlas the circle is a *differentiable manifold*. It is also *smooth* and *analytic* because the transition functions have these properties as well. Other circle properties allow it to meet the requirements of more specialised types of manifold. For example, the circle has a notion of distance between two points, the arc-length between the points; hence it is a *Riemannian manifold*.

History

The study of manifolds combines many important areas of mathematics: it generalises concepts such as curves and surfaces as well as ideas from linear algebra and topology.

Early Development

Before the modern concept of a manifold there were several important results.

Non-Euclidean geometry considers spaces where Euclid's parallel postulate fails. Saccheri first studied them in 1733.

Lobachevsky, Bolyai, and Riemann developed them 100 years later. Their research uncovered two types of spaces whose geometric structures differ from that of classical Euclidean space; these gave rise to hyperbolic geometry and elliptic geometry. In the modern theory of manifolds, these notions correspond to Riemannian manifolds with constant negative and positive curvature, respectively.

Carl Friedrich Gauss may have been the first to consider abstract spaces as mathematical objects in their own right. His theorema egregium gives a method for computing the curvature of a surface without considering the ambient space in which the surface lies.

Such a surface would, in modern terminology, be called a manifold; and in modern terms, the theorem proved that the curvature of the surface is an intrinsic property. Manifold theory has come to focus exclusively on these intrinsic properties (or invariants), while largely ignoring the extrinsic properties of the ambient space.

Another, more topological example of an intrinsic property of a manifold is its Euler characteristic. Leonhard Euler showed that for a convex polytope in the three-dimensional Euclidean space with V vertices (or corners), E edges, and F faces,

$$V - E + F = 2.$$

The same formula will hold if we project the vertices and edges of the polytope onto a sphere, creating a topological map with V vertices, E edges, and F faces, and in fact, will remain true for any spherical map, even if it does not arise from any convex polytope.

Thus 2 is a topological invariant of the sphere, called its Euler characteristic. On the other hand, a torus can be sliced open by its 'parallel' and 'meridian' circles, creating a map with $V = 1$ vertex, $E = 2$ edges, and $F = 1$ face.

Thus the Euler characteristic of the torus is $1 - 2 + 1 = 0$. The Euler characteristic of other surfaces is a useful topological invariant, which can be extended to higher dimensions using Betti numbers. In the mid nineteenth century, the Gauss–Bonnet theorem linked the Euler characteristic to the Gaussian curvature.

Synthesis

Investigations of Niels Henrik Abel and Carl Gustav Jacobi on inversion of elliptic integrals in the first half of 19th century led them to consider special types of complex manifolds, now known as Jacobians. Bernhard Riemann further contributed to their theory, clarifying the geometric meaning of the process of analytic continuation of functions of complex variables.

Another important source of manifolds in 19th century mathematics was analytical mechanics, as developed by Simeon Poisson, Jacobi, and William Rowan Hamilton. The possible states of a mechanical system are thought to be points of an abstract space, phase space in Lagrangian and Hamiltonian formalisms of classical mechanics. This space is, in fact, a high-dimensional manifold, whose dimension corresponds to the degrees of freedom of the system and where the points are specified by their generalised coordinates.

For an unconstrained movement of free particles the manifold is equivalent to the Euclidean space, but various conservation laws constrain it to more complicated formations, e.g. Liouville tori. The theory of a rotating solid body, developed in the 18th century by Leonhard Euler and Joseph Lagrange, gives another example where the manifold is nontrivial. Geometrical and topological aspects of classical mechanics were emphasised by Henri Poincaré, one of the founders of topology.

Riemann was the first one to do extensive work generalising the idea of a surface to higher dimensions. The name *manifold* comes from Riemann's original German term, *Mannigfaltigkeit*, which William Kingdon Clifford translated as "manifoldness". In his Göttingen inaugural lecture, Riemann described the set of all possible values of a variable with certain constraints as a *Mannigfaltigkeit*, because the variable can have *many* values. He distinguishes between *stetige Mannigfaltigkeit* and *diskrete Mannigfaltigkeit* (*continuous manifoldness* and *discontinuous manifoldness*), depending on whether the value changes continuously or not. As continuous examples, Riemann refers to not only colours and the locations of objects in space, but also the possible shapes of a spatial figure. Using induction, Riemann constructs an *n-fach ausgedehnte Mannigfaltigkeit* (*n times extended manifoldness* or *n-dimensional manifoldness*) as a continuous stack of (n–1) dimensional manifoldnesses. Riemann's intuitive notion of a *Mannig-faltigkeit* evolved into what is today formalised as a manifold. Riemannian manifolds and Riemann surfaces are named after Riemann.

Topology of Manifolds: Highlights

Two-dimensional manifolds, also known as a 2D *surfaces* embedded in our common 3D space, were considered by Riemann under the guise of Riemann surfaces, and rigorously classified in the beginning of the 20th century by Poul Heegaard and Max Dehn. Henri Poincaré pioneered the study of three-dimensional manifolds and raised a fundamental question about them, today known as the Poincaré conjecture.

After nearly a century of effort by many mathematicians, starting with Poincaré himself, a consensus among experts (as of 2006) is that Grigori Perelman has proved the Poincaré conjecture. Bill Thurston's geometrisation programme, formulated in the 1970s, provided a far-reaching extension of the Poincaré conjecture to the general three-dimensional manifolds. Four-dimensional manifolds were brought to the forefront of mathematical research in the 1980s by Michael Freedman and in a different setting, by Simon Donaldson, who was motivated by the then recent progress in theoretical physics (Yang-Mills theory), where they serve as a substitute for ordinary 'flat' space-time. Andrey Markov Jr. showed in 1960 that no algorithm exists for classifying four-dimensional manifolds. Important work on higher-dimensional manifolds, including analogues of the Poincaré conjecture, had been done earlier by René Thom, John Milnor, Stephen Smale and Sergei Novikov. One of the most pervasive and flexible techniques underlying much work on the topology of manifolds is Morse theory.

Mathematical definition

Informally, a manifold is a space that is "modeled on" Euclidean space.

There are many different kinds of manifolds and generalisations. In geometry and topology, all manifolds are topological manifolds, possibly with additional structure, most often a differentiable structure. In terms of constructing manifolds via patching, a manifold has an additional structure if the transition maps between different patches satisfy axioms beyond just continuity. For instance, differentiable manifolds have homeomorphisms on overlapping neighbourhoods diffeomorphic with each other, so that the manifold has a well-defined set of functions which are differentiable in each neighbourhood, and so differentiable on the manifold as a whole.

Formally, a topological manifold is a second countable Hausdorff space that is locally homeomorphic to Euclidean space.

Second countable and *Hausdorff* are point-set conditions; *second countable* excludes spaces which are in some sense 'too large' such as the long line, while *Hausdorff* excludes spaces such as "the line with two origins" (these generalisations of manifolds are discussed in non-Hausdorff manifolds).

Locally homeomorphic to Euclidean space means that every point has a neighbourhood homeomorphic to an open Euclidean n-ball,

$$\mathbf{B}^n = \{(x_1, x_2, \ldots, x_n) \in \mathbb{R}^n \; x_1^2 + x_2^2 + \ldots + x_n^2 < 1\}.$$

Generally manifolds are taken to have a fixed dimension (the space must be locally homeomorphic to a fixed n-ball), and such a space is called an n-manifold; however, some authors admit manifolds where different points can have different dimensions. If a manifold has a fixed dimension, it is called a pure manifold. For example, the sphere has a constant dimension of 2 and is therefore a pure manifold whereas the disjoint union of a sphere and a line in three-dimensional space is *not* a pure manifold. Since dimension is a local invariant (i.e. the map sending each point to the dimension of its neighbourhood over which a chart is defined, is locally constant), each connected component has a fixed dimension. Scheme-theoretically, a manifold is a locally ringed space, whose structure sheaf is locally isomorphic to the sheaf of continuous (or differentiable, or complex-analytic, etc.) functions on Euclidean space. This definition is mostly used when discussing analytic manifolds in algebraic geometry.

Broad Definition

The broadest common definition of manifold is a topological space locally homeomorphic to a topological vector space over the reals. This omits the point-set axioms, allowing higher cardinalities and non-Hausdorff manifolds; and it omits finite dimension, allowing structures such as Hilbert manifolds to be modeled on Hilbert spaces, Banach manifolds to be modeled on Banach spaces, and Fréchet manifolds to be modeled on Fréchet spaces. Usually one relaxes one or the other condition: manifolds with the point-set axioms are studied in general topology, while infinite-dimensional manifolds are studied in functional analysis.

Charts, Atlases, and Transition Maps

The spherical Earth is navigated using flat maps or charts, collected in an atlas. Similarly, a differentiable manifold can be described using mathematical maps, called *coordinate charts*, collected in a mathematical *atlas*. It is not generally possible to describe a manifold with just one chart, because the global structure of the manifold is different from the simple structure of the charts. For example, no single flat map can represent the entire Earth without separation of adjacent features across the map's boundaries or duplication of coverage. When a manifold is constructed from multiple overlapping charts, the regions where they overlap carry information essential to understanding the global structure.

Charts

A coordinate map, a coordinate chart, or simply a chart, of a manifold is an invertible map between a subset of the manifold and a simple space such that both the map and its inverse preserve the desired structure. For a topological manifold, the simple space is some Euclidean space R^n and interest focuses on the topological structure. This structure is preserved by homeomorphisms, invertible maps that are continuous in both directions.

In the case of a differentiable manifold, a set of charts called an atlas allows us to do calculus on manifolds. Polar coordinates, for example, form a chart for the plane R^2 minus the positive x-axis and the origin. Another example of a chart is the map χ_{top} mentioned in the section above, a chart for the circle.

Atlases

The description of most manifolds requires more than one chart (a single chart is adequate for only the simplest manifolds). A specific collection of charts which covers a manifold is called an atlas. An atlas is not unique as all manifolds can be covered multiple ways using different combinations of charts. Two atlases are said to be C^k equivalent if their union is also a C^k atlas. The atlas containing all possible charts consistent with a given atlas is called the maximal atlas (i.e. an equivalence class containing that given atlas. Unlike an ordinary atlas, the maximal atlas of a given manifold is unique. Though it is useful for definitions, it is an abstract object and not used directly (e.g. in calculations).

Transition Maps

Charts in an atlas may overlap and a single point of a manifold may be represented in several charts. If two charts overlap, parts of them represent the same region of the manifold, just as a map of Europe and a map of Asia may both contain Moscow. Given two overlapping charts, a transition function can be defined which goes from an open ball in R^n to the manifold and then back to another (or perhaps the same) open ball in R^n. The resultant map, like the map T in the circle example above, is called a change of coordinates, a coordinate transformation, a transition function, or a transition map.

Additional Structure

An atlas can also be used to define additional structure on the manifold. The structure is first defined on each chart separately. If all the transition maps are compatible with this structure, the structure transfers to the manifold. This is the standard way differentiable manifolds are defined. If the transition functions of an atlas for a topological manifold preserve the natural differential structure of R^n (that is, if they are diffeomorphisms), the differential structure transfers to the manifold and turns it into a differentiable manifold. Complex manifolds are introduced in an analogous way by requiring that the transition functions of an atlas are holomorphic functions. For symplectic manifolds, the transition functions must be symplectomorphisms. The structure on the manifold depends on the atlas, but sometimes different atlases can be said to give rise to the same structure. Such atlases are called compatible. These notions are made precise in general through the use of pseudogroups.

Construction

A single manifold can be constructed in different ways, each stressing a different aspect of the manifold, thereby leading to a slightly different viewpoint.

Charts

Figure: *The chart maps the part of the sphere with positive z coordinate to a disc.*

Perhaps the simplest way to construct a manifold is the one used in the example above of the circle. First, a subset of R^2 is identified, and then an atlas covering this subset is constructed. The concept of *manifold* grew historically from constructions like this. Here is another example, applying this method to the construction of a sphere:

Sphere with Charts

A sphere can be treated in almost the same way as the circle. In mathematics a sphere is just the surface (not the solid interior), which can be defined as a subset of R^3:

$$S = \{(x, y, z) \in \mathbf{R}^3 \mid x^2 + y^2 + z^2 = 1\}.$$

The sphere is two-dimensional, so each chart will map part of the sphere to an open subset of R^2. Consider the northern hemisphere, which is the part with positive z coordinate (coloured red in the picture on the right). The function χ defined by

$$\chi(x, y, z) = (x, y),$$

maps the northern hemisphere to the open unit disc by projecting it on the (x, y) plane. A similar chart exists for the southern hemisphere. Together with two charts projecting on the (x, z) plane and two charts projecting on the (y, z) plane, an atlas of six charts is obtained which covers the entire sphere.

This can be easily generalised to higher-dimensional spheres.

Patchwork

A manifold can be constructed by gluing together pieces in a consistent manner, making them into overlapping charts.

This construction is possible for any manifold and hence it is often used as a characterisation, especially for differentiable and Riemannian manifolds. It focuses on an atlas, as the patches naturally provide charts, and since there is no exterior space involved it leads to an intrinsic view of the manifold.

The manifold is constructed by specifying an atlas, which is itself defined by transition maps. A point of the manifold is therefore an equivalence class of points which are mapped to each other by transition maps. Charts map equivalence classes to points of a single patch. There are usually strong demands on the consistency of the transition maps. For topological manifolds they are required to be homeomorphisms; if they are also diffeomorphisms, the resulting manifold is a differentiable manifold.

This can be illustrated with the transition map $t = {}^1D_s$ from the second half of the circle example. Start with two copies of the line. Use the coordinate s for the first copy, and t for the second copy. Now, glue both copies together by identifying the point t on the second copy with the point $s = {}^1D_t$ on the first copy (the points $t = 0$ and $s = 0$ are not identified with any point on the first and second copy, respectively). This gives a circle.

Intrinsic and Extrinsic View

The first construction and this construction are very similar, but they represent rather different points of view. In the first construction, the manifold is seen as embedded in some Euclidean space. This is the *extrinsic view*. When a manifold is viewed in this way, it is easy to use intuition from Euclidean spaces to define additional structure. For example, in a Euclidean space it is always clear whether a vector at some point is tangential or normal to some surface through that point.

The patchwork construction does not use any embedding, but simply views the manifold as a topological space by itself. This abstract point of view is called the *intrinsic view*. It can make it harder to imagine what a tangent vector might be, and there is no intrinsic notion of a normal bundle, but instead there is an intrinsic stable normal bundle.

n-Sphere as a Patchwork

The n-sphere S^n is a generalisation of the idea of a circle (1-sphere) and sphere (2-sphere) to higher dimensions. An n-sphere S^n can be constructed by gluing together two copies of R^n. The transition map between them is defined as

$$\mathbf{R}^n \setminus \{0\} \to \mathbf{R}^n \setminus \{0\} : x \mapsto x / \| x \|^2 .$$

This function is its own inverse and thus can be used in both directions. As the transition map is a smooth function, this atlas defines a smooth manifold. In the case $n = 1$, the example simplifies to the circle example given earlier.

Identifying Points of a Manifold

It is possible to define different points of a manifold to be same. This can be visualised as gluing these points together in a single point, forming a quotient space. There is, however, no reason to expect such quotient spaces to be manifolds. Among the possible quotient spaces that are not necessarily manifolds, orbifolds and CW complexes are considered to be relatively well-behaved. An example of a quotient space of a manifold that is also a manifold is the real projective space identified as a quotient space of the corresponding sphere.

One method of identifying points (gluing them together) is through a right (or left) action of a group, which acts on the manifold. Two points are identified if one is moved onto the other by some group element. If M is the manifold and G is the group, the resulting quotient space is denoted by M / G (or $G \setminus M$).

Manifolds which can be constructed by identifying points include tori and real projective spaces (starting with a plane and a sphere, respectively).

Gluing Along Boundaries

Two manifolds with boundaries can be glued together along a boundary. If this is done the right way, the result is also a manifold. Similarly, two boundaries of a single manifold can be glued together.

Formally, the gluing is defined by a bijection between the two boundaries. Two points are identified when they are mapped onto each other. For a topological manifold this bijection should be a homeomorphism, otherwise the result will not be a topological manifold. Similarly for a differentiable manifold it has to be a diffeomorphism. For other manifolds other structures should be preserved.

A finite cylinder may be constructed as a manifold by starting with a strip $[0, 1] \times [0, 1]$ and gluing a pair of opposite edges on the boundary by a suitable diffeomorphism. A projective plane may be obtained by gluing a sphere with a hole in it to a Möbius strip along their respective circular boundaries.

Manifolds with additional structure

Topological Manifolds

The simplest kind of manifold to define is the topological manifold, which looks locally like some "ordinary" Euclidean space R^n. Formally, a topological manifold is a topological space locally homeomorphic to a Euclidean space. This means that every point has a neighbourhood for which there exists a homeomorphism (a bijective continuous function whose inverse is also continuous) mapping that neighbourhood to R^n. These homeomorphisms are the charts of the manifold.

It is to be noted that a *topological* manifold looks locally like a Euclidean space in a rather weak manner: while for each individual chart it is possible to distinguish differentiable functions or measure distances and angles, merely by virtue of being a topological manifold a space does not have any *particular* and *consistent* choice of such concepts. In order to discuss such properties for a manifold, one needs to specify further structure and consider differentiable manifolds and Riemannian manifolds discussed below. In particular, the same underlying topological manifold can have several mutually incompatible classes of differentiable functions and an infinite number of ways to specify distances and angles.

Usually additional technical assumptions on the topological space are made to exclude pathological cases. It is customary to require that the space be Hausdorff and second countable.

The *dimension* of the manifold at a certain point is the dimension of the Euclidean space that the charts at that point map to (number n in the definition). All points in a connected manifold have the same dimension. Some authors require that all charts of a topological manifold map to Euclidean spaces of same dimension. In that case every topological manifold has a topological invariant, its dimension. Other authors allow disjoint unions of topological manifolds with differing dimensions to be called manifolds.

Differentiable Manifolds

For most applications a special kind of topological manifold, a differentiable manifold, is used. If the local charts on a manifold are compatible in a certain sense, one can define directions, tangent spaces, and differentiable functions on that manifold. In particular it is possible to use calculus on a differentiable manifold. Each point of an n-dimensional differentiable manifold has a tangent space. This is an n-dimensional Euclidean space consisting of the tangent vectors of the curves through the point.

Two important classes of differentiable manifolds are smooth and analytic manifolds. For smooth manifolds the transition maps are smooth, that is infinitely differentiable. Analytic manifolds are smooth manifolds with the additional condition that the transition maps are analytic (they can be expressed as power series). The sphere can be given analytic structure, as can most familiar curves and surfaces.

There are also topological manifolds, i.e., locally Euclidean spaces, which possess no differentiable structures at all. A rectifiable set generalises the idea of a piecewise smooth or rectifiable curve to higher dimensions; however, rectifiable sets are not in general manifolds.

Riemannian Manifolds

To measure distances and angles on manifolds, the manifold must be Riemannian. A Riemannian manifold is a differentiable manifold in which each tangent space is equipped with an inner product 0Å–, Å– 0 in a manner which varies smoothly from point to point. Given two tangent vectors u and v, the inner product0u,v 0gives a real number. The dot (or scalar) product is a typical example of an inner product. This allows one to define various notions such as length, angles, areas (or volumes), curvature, gradients of functions and divergence of vector fields.

All differentiable manifolds (of constant dimension) can be given the structure of a Riemannian manifold. The Euclidean space itself carries a natural structure of Riemannian manifold (the tangent spaces are naturally identified with the Euclidean space itself and carry the standard scalar product of the space). Many familiar curves and surfaces, including for example all n-spheres, are specified as subspaces of a Euclidean space and inherit a metric from their embedding in it.

Finsler Manifolds

A Finsler manifold allows the definition of distance but does not require the concept of angle; it is an analytic manifold in which each tangent space is equipped with a norm, $||\cdot||$, in a manner which varies smoothly from point to point. This norm can be extended to a metric, defining the length of a curve; but it cannot in general be used to define an inner product. Any Riemannian manifold is a Finsler manifold.

Lie Groups

Lie groups, named after Sophus Lie, are differentiable manifolds that carry also the structure of a group which is such that the group operations are defined by smooth maps.

A Euclidean vector space with the group operation of vector addition is an example of a non-compact Lie group. A simple example of a compact Lie group is the circle: the group operation is simply rotation. This group, known as U(1), can be also characterised as the group of complex numbers of modulus 1 with multiplication as the group operation. Other examples of Lie groups include special groups of matrices, which are all subgroups of the general linear group, the group of n by n matrices with non-zero determinant. If the matrix entries are real numbers, this will be an n^2-dimensional disconnected manifold. The orthogonal groups, the symmetry groups of the sphere and hyperspheres, are $n(n-1)/2$ dimensional manifolds, where n-1 is the dimension of the sphere. Further examples can be found in the table of Lie groups.

Other Types of Manifolds

- A complex manifold is a manifold modeled on C^n with holomorphic transition functions on chart overlaps. These manifolds are the basic objects of study in complex geometry. A one-complex-dimensional manifold is called a Riemann surface. Note that an n-dimensional complex manifold has dimension $2n$ as a real differentiable manifold.
- A CR manifold is a manifold modeled on boundaries of domains in C^n.
- Infinite dimensional manifolds: to allow for infinite dimensions, one may consider Banach manifolds which are locally homeomorphic to Banach spaces. Similarly, Fréchet manifolds are locally homeomorphic to Fréchet spaces.
- A symplectic manifold is a kind of manifold which is used to represent the phase spaces in classical mechanics. They are endowed with a 2-form that defines the Poisson bracket. A closely related type of manifold is a contact manifold.
- A combinatorial manifold is a kind of manifold which is discretisation of a manifold. It usually means a piecewise linear manifold made by simplicial complexes.
- A digital manifold is a special kind of combinatorial manifold which is defined in digital space.

Classification and Invariants

Different notions of manifolds have different notions of classification and invariant; in this section we focus on smooth closed manifolds.

The classification of smooth closed manifolds is well-understood *in principle*, except in dimension 4: in low dimensions (2 and 3) it is geometric, via the uniformisation theorem and the Solution of the Poincaré conjecture, and in high dimension (5 and above) it is algebraic, via surgery theory. This is a classification in principle: the general question of whether two smooth manifolds are diffeomorphic is not computable in general. Further, specific computations remain difficult, and there are many open questions.

Orientable surfaces can be visualised, and their diffeomorphism classes enumerated, by genus. Given two orientable surfaces, one can determine if they are diffeomorphic by computing their respective genera and comparing: they are diffeomorphic if and only if the genera are equal, so the genus forms a complete set of invariants.

This is much harder in higher dimensions: higher dimensional manifolds cannot be directly visualised (though visual intuition is useful in understanding them), nor can their diffeomorphism classes be enumerated, nor can one in general determine if two different descriptions of a higher-dimensional manifold refer to the same object.

However, one can determine if two manifolds are *different* if there is some intrinsic characteristic that differentiates them.

Such criteria are commonly referred to as invariants, because, while they may be defined in terms of some presentation (such as the genus in terms of a triangulation), they are the same relative to all possible descriptions of a particular manifold: they are *invariant* under different descriptions.

Naively, one could hope to develop an arsenal of invariant criteria that would definitively classify all manifolds up to isomorphism. Unfortunately, it is known that for manifolds of dimension 4 and higher, no programme exists that can decide whether two manifolds are diffeomorphic.

Smooth manifolds have a rich set of invariants, coming from point-set topology, classic algebraic topology, and geometric topology. The most familiar invariants, which are visible for surfaces, are orientability (a normal invariant, also detected by homology) and genus (a homological invariant).

Smooth closed manifolds have no local invariants (other than dimension), though geometric manifolds have local invariants, notably the curvature of a Riemannian manifold and the torsion of a manifold equipped with an affine connection. This distinction between no local invariants and local invariants is a common way to distinguish between geometry and topology. All invariants of a smooth closed manifold are thus global.

Algebraic topology is a source of a number of important global invariant properties. Some key criteria include the *simply connected* property and orientability. Indeed several branches of mathematics, such as homology and homotopy theory, and the theory of characteristic classes were founded in order to study invariant properties of manifolds.

Genus and the Euler Characteristic

For two dimensional manifolds a key invariant property is the genus, or the "number of handles" present in a surface. A torus is a sphere with one handle, a double torus is a sphere with two handles, and so on. Indeed it is possible to fully characterise compact, two-dimensional manifolds on the basis of genus and orientability. In higher-dimensional manifolds genus is replaced by the notion of Euler characteristic, and more generally Betti numbers and homology and cohomology.

Maps of Manifolds

Just as there are various types of manifolds, there are various types of maps of manifolds. In addition to continuous functions and smooth functions generally, there are maps with special properties. In geometric topology a basic type are embeddings, of which knot theory is a central example, and generalisations such as immersions, submersions, covering spaces, and ramified covering spaces. Basic results include the Whitney embedding theorem and Whitney immersion theorem.

In Riemannian geometry, one may ask for maps to preserve the Riemannian metric, leading to notions of isometric embeddings, isometric immersions, and Riemannian submersions; a basic result is the Nash embedding theorem.

Generalisations of Manifolds

- *Orbifolds:* An orbifold is a generalisation of manifold allowing for certain kinds of "singularities" in the topology. Roughly speaking, it is a space which locally looks like the quotients of some simple space (*e.g.* Euclidean space) by the actions of various finite groups. The singularities correspond to fixed points of the group actions, and the actions must be compatible in a certain sense.
- Manifold with corners
- *Algebraic varieties and schemes:* Non-singular algebraic varieties over the real or complex numbers are manifolds. One generalises this first by allowing singularities, secondly by allowing different fields, and thirdly by emulating the patching construction of manifolds: just as a manifold is glued together from open subsets of Euclidean space, an algebraic variety is glued together from affine algebraic varieties, which are zero sets of polynomials over algebraically closed fields. Schemes

are likewise glued together from affine schemes, which are a generalisation of algebraic varieties. Both are related to manifolds, but are constructed algebraically using sheaves instead of atlases.

Because of singular points, a variety is in general not a manifold, though linguistically the French *variété*, German *Mannigfaltigkeit* and English *manifold* are largely synonymous. In French an algebraic variety is called *une variété algébrique* (an *algebraic variety*), while a smooth manifold is called *une variété différentielle* (a *differential variety*).

- *CW-complexes:* A CW complex is a topological space formed by gluing disks of different dimensionality together. In general the resulting space is singular, and hence not a manifold. However, they are of central interest in algebraic topology, especially in homotopy theory, as they are easy to compute with and singularities are not a concern.

Change of Coordinates

In geometry and kinematics, coordinate systems are used not only to describe the (linear) position of points, but also to describe the angular position of axes, planes, and rigid bodies.

In the latter case, the orientation of a second (typically referred to as "local") coordinate system, fixed to the node, is defined based on the first (typically referred to as "global" or "world" coordinate system).

For instance, the orientation of a rigid body can be represented by an orientation matrix, which includes, in its three columns, the Cartesian coordinates of three points. These points are used to define the orientation of the axes of the local system; they are the tips of three unit vectors aligned with those axes.

Transformations

A coordinate transformation is a conversion from one system to another, to describe the same space.

With every bijection from the space to itself two coordinate transformations can be associated:

- Such that the new coordinates of the image of each point are the same as the old coordinates of the original point (the formulas for the mapping are the inverse of those for the coordinate transformation)

- Such that the old coordinates of the image of each point are the same as the new coordinates of the original point (the formulas for the mapping are the same as those for the coordinate transformation)

For example, in 1D, if the mapping is a translation of 3 to the right, the first moves the origin from 0 to 3, so that the coordinate of each point becomes 3 less, while the second moves the origin from 0 to -3, so that the coordinate of each point becomes 3 more.

Systems Commonly Used

Some coordinate systems are the following:

- The Cartesian coordinate system (also called the "rectangular coordinate system"), which, for two- and three-dimensional spaces, uses two and three numbers (respectively) representing distances from the origin in three mutually perpendicular directions.
- Curvilinear coordinates are a generalisation of coordinate systems generally; the system is based on the intersection of curves.
- Polar coordinate system represents a point in the plane by a distance from the origin and an angle measured from a reference line intersecting the origin.
- Log-polar coordinate system represents a point in the plane by the logarithm of the distance from the origin and an angle measured from a reference line intersecting the origin.
- Cylindrical coordinate system represents a point in three-space using two perpendicular axes; distance is measured along one axis, while the other axis formes the reference line for a polar coordinate representation of the remaining two components.
- Spherical coordinate system represents a point in three space by the distance from the origin and two angles measured from two reference lines which intersect the origin.
- Plücker coordinates are a way of representing lines in 3D Euclidean space using a six-tuple of numbers as homogeneous coordinates.
- Generalised coordinates are used in the Lagrangian treatment of mechanics.

- Canonical coordinates are used in the Hamiltonian treatment of mechanics.
- Parallel coordinates visualise a point in n-dimensional space as a polyline connecting points on n vertical lines.
- Barycentric coordinates (mathematics) as used for Ternary_plot

There are ways of describing curves without coordinates, using intrinsic equations that use invariant quantities such as curvature and arc length. These include:

- Whewell equation relates arc length and tangential angle.
- Cesàro equation relates arc length and curvature.

Chapter

3

6-Sphere Coordinates

In mathematics, 6-sphere coordinates are the coordinate system created by inverting the Cartesian coordinates across the unit sphere. The three coordinates are

$$u = \frac{x}{x^2 + y^2 + z^2}, \quad v = \frac{y}{x^2 + y^2 + z^2}, \quad w = \frac{z}{x^2 + y^2 + z^2}.$$

Since inversion is its own inverse, the equations for x, y, and z in terms of u, v, and w are similar:

$$x = \frac{u}{u^2 + v^2 + w^2}, \quad y = \frac{v}{u^2 + v^2 + w^2}, \quad z = \frac{w}{u^2 + v^2 + w^2}.$$

This coordinate system is R-separable for the 3-variable Laplace equation.

Abscissa

In mathematics, abscissa (plural abscissae or abscissæ) refers to that element of an ordered pair which is plotted on the horizontal axis of a two-dimensional Cartesian coordinate system, as opposed to the ordinate. It is the first of the two terms (often labelled x and y) which define the location of a point in such a coordinate system.

$$(\overbrace{x}^{\text{abscissa}} , \overbrace{y}^{\text{ordinate}})$$

The usage of the word abscissa is first recorded in 1659 by Stefano degli Angeli, a mathematics professor in Rome, according to Moritz Cantor. Soon thereafter, Leibniz used the term extensively in Latin in his *Mathematische Schriften* (1692), after which it became a standardised mathematical term. The first occurrence of the term in English is found in *An Institution of Fluxions* by the English mathematician Humphry Ditton (1706), where he spells the word *abscisse*, possibly denoting the plural.

Examples

For the point (-7, 3), -7 is called the abscissa and 3 the ordinate.

Affine Coordinate System

In mathematics, an affine coordinate system is a coordinate system on an affine space where each coordinate is an affine map to the number line. In other words, it is an injective affine map from an affine space A to the coordinate space k^n, where k is the base field, for example, the real numbers R. The most important case of affine coordinates in Euclidean spaces is real-valued Cartesian coordinate system. Non-orthogonal affine coordinates are referred to as oblique.

A system of n coordinates on n-dimensional space is defined by a $(n+1)$-tuple $(O, R_1, \ldots R_n)$ of points not belonging to any affine subspace of a lesser dimension. Any given coordinate n-tuple gives the point by the formula:

$$(x_1, \ldots x_n) \mapsto O + x_1 (R_1 - O) + \ldots + x_n (R_n - O).$$

Note that $R_j - O$ are *difference* vectors with the origin in O and ends in R_j.

An affine space cannot have a coordinate system with n less than its dimension, but n may indeed be greater, which means that the coordinate map is not necessary surjective. Examples of n-coordinate system in an $(n-1)$-dimensional space are barycentric coordinates and affine homogeneous coordinates $(1, x_1, \ldots x_{n-1})$. In the latter case the x_0 coordinate is equal to 1 on all space, but this "reserved" coordinate facilitates matrix representation of affine maps.

Alpha-numeric Grid

	a	*b*	*c*	*d*	*e*	*f*
1	a1	b1	c1	d1	e1	f1
2	a2	b2	c2	d2	e2	f2
3	a3	b3	c3	d3	e3	f3
4	a4	b4	c4	d4	e4	f4
5	a5	b5	c5	d5	e5	f5
6	a6	b6	c6	d6	e6	f6

An alphanumeric grid (also known as atlas grid) is a simple coordinate system on a grid in which each cell is identified by a combination of a letter and a number. An advantage over numeric coordinates, which use two numbers instead of a number and a letter to refer to a grid cell, is that there can be no confusion over which coordinate refers to which direction. Algebraic chess notation uses an alphanumeric grid to refer to the squares of a chessboard.

Astronomical Coordinate Systems

Astronomical coordinate systems are coordinate systems used in astronomy to describe the location of objects in the sky and in the universe. The most commonly occurring such systems are coordinate systems on the celestial sphere, but extragalactic coordinates systems are also important for describing more distant objects.

Coordinate Systems on the Celestial Sphere

- Horizontal coordinate system
- Equatorial coordinate system - based on Earth rotation
- Ecliptic coordinate system - based on Solar System rotation
- Galactic coordinate system - based on Milky Way rotation

Bipolar Coordinates

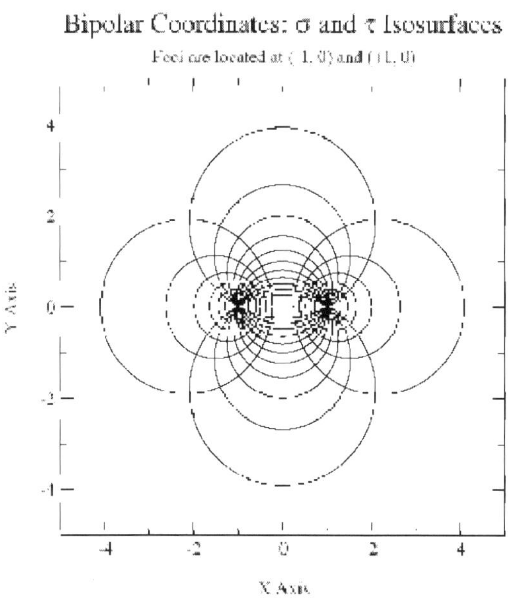

Figure: *Bipolar coordinate system*

Bipolar coordinates are a two-dimensional orthogonal coordinate system. There are two commonly defined types of bipolar coordinates. The other system is two-centre bipolar coordinates. There is also a third coordinate system that is based on two poles (biangular coordinates). The first is based on the Apollonian circles. The curves of constant σ and of τ are circles that intersect at right angles. The coordinates have two foci F_1 and F_2, which are generally taken to be fixed at $(-a, 0)$ and $(a, 0)$, respectively, on the x-axis of a Cartesian coordinate system. Bipolar coordinates form the basis for several sets of three-dimensional orthogonal coordinates. The bipolar cylindrical coordinates are produced by projecting in the z-direction. The bispherical coordinates are produced by rotating the bipolar coordinates about the x-axis, i.e., the axis connecting the foci, whereas the toroidal coordinates are produced by rotating the bipolar coordinates about the y-axis, i.e., the axis separating the foci. The classic applications of bipolar coordinates are in solving partial differential equations, e.g., Laplace's equation or the Helmholtz equation, for which bipolar coordinates allow a separation of variables. A typical example would be the electric field surrounding two parallel cylindrical conductors. The term "bipolar" is sometimes used to describe other curves having two singular points (foci), such as ellipses, hyperbolas, and Cassini ovals. However, the term *bipolar coordinates* is reserved for the coordinates described here, and never used to describe coordinates associated with those other curves, such as elliptic coordinates.

Definition

The most common definition of bipolar coordinates (σ, τ) is

$$x = a\frac{\sinh\tau}{\cosh\tau - \cos\sigma}$$

$$y = a\frac{\sin\sigma}{\cosh\tau - \cos\sigma}$$

where the σ-coordinate of a point P equals the angle $F_1 P F_2$ and the τ-coordinate equals the natural logarithm of the ratio of the distances d_1 and d_2 to the foci

$$\tau = \ln\frac{d_1}{d_2}$$

(Recall that F_1 and F_2 are located at $(-a, 0)$ and $(a, 0)$, respectively.) Equivalently

$$x + iy = ai\cot\left(\frac{\sigma + i\tau}{2}\right)$$

Curves of Constant σ and τ

Scale Factors

The scale factors for the bipolar coordinates (σ, τ) are equal

$$h_\sigma = h_\tau = \frac{a}{\cosh\tau - \cos\sigma}$$

Thus, the infinitesimal area element equals

$$dA = \frac{a^2}{\left(\cosh\tau - \cos\sigma\right)^2} d\sigma d\tau$$

and the Laplacian is given by

$$\nabla^2\Phi = \frac{1}{a^2}\left(\cosh\tau - \cos\sigma\right)^2 \left(\frac{\partial^2\Phi}{\partial\sigma^2} + \frac{\partial^2\Phi}{\partial\tau^2}\right)$$

Other differential operators such as $\nabla \cdot \mathbf{F}$ and $\nabla \times \mathbf{F}$ can be expressed in the coordinates (σ, τ) by substituting the scale factors into the general formulae found in orthogonal coordinates.

Bipolar Cylindrical Coordinates

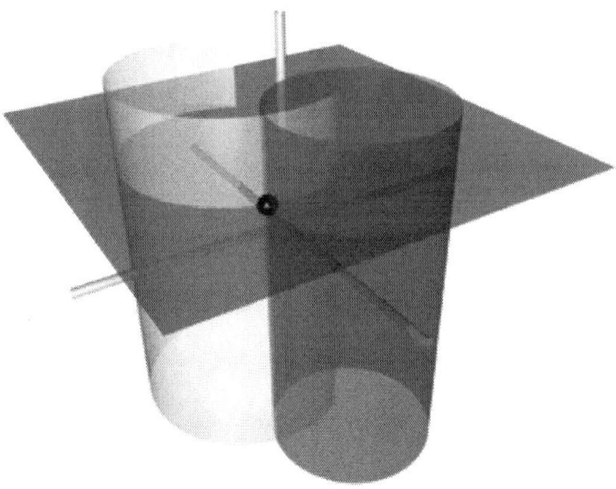

Figure: *Coordinate surfaces of the bipolar cylindrical coordinates. The yellow crescent corresponds to σ, whereas the red tube corresponds to τ and the blue plane corresponds to z = 1. The three surfaces intersect at the point P (shown as a black sphere).*

Bipolar cylindrical coordinates are a three-dimensional orthogonal coordinate system that results from projecting the two-dimensional bipolar coordinate system in the perpendicular z-direction. The two lines of foci F_1 and F_2 of the projected Apollonian circles are generally taken to be defined by $x = -a$ and $x = +a$, respectively, (and by $y = 0$) in the Cartesian coordinate system. The term "bipolar" is often used to describe other curves having two singular points (foci), such as ellipses, hyperbolas, and Cassini ovals. However, the term *bipolar coordinates* is never used to describe coordinates associated with those curves, e.g., elliptic coordinates.

Basic Definition

The most common definition of bipolar cylindrical coordinates (σ, τ, z) is

$$x = a\frac{\sinh\tau}{\cosh\tau - \cos\sigma}$$

$$y = a\frac{\sin\sigma}{\cosh\tau - \cos\sigma}$$

$$z = z$$

where the σ coordinate of a point P equals the angle F_1PF_2 and the τ coordinate equals the natural logarithm of the ratio of the distances d_1 and d_2 to the focal lines

$$\tau = \ln\frac{d_1}{d_2}$$

(Recall that the focal lines F_1 and F_2 are located at $x = -a$ and $x = +a$, respectively.)

Surfaces of constant σ correspond to cylinders of different radii

$$x^2 + (y - a\cot\sigma)^2 = \frac{a^2}{\sin^2\sigma}$$

that all pass through the focal lines and are not concentric. The surfaces of constant τ are non-intersecting cylinders of different radii

$$y^2 + (x - a\coth\tau)^2 = \frac{a^2}{\sinh^2\tau}$$

that surround the focal lines but again are not concentric. The focal lines and all these cylinders are parallel to the z-axis (the direction of projection). In the $z = 0$ plane, the centres of the constant-σ and constant- cylinders lie on the and axes, respectively.

Scale Factors

The scale factors for the bipolar coordinates σ and τ are equal

$$h_\sigma = h_\tau = \frac{a}{\cosh\tau - \cos\sigma}$$

whereas the remaining scale factor $h_z = 1$. Thus, the infinitesimal volume element equals

$$dV = \frac{a^2}{(\cosh\tau - \cos\sigma)^2} d\sigma d\tau dz$$

and the Laplacian is given by

$$\nabla^2\Phi = \frac{1}{a^2}(\cosh\tau - \cos\sigma)^2 \left(\frac{\partial^2\Phi}{\partial\sigma^2} + \frac{\partial^2\Phi}{\partial\tau^2} \right) + \frac{\partial^2\Phi}{\partial z^2}$$

Other differential operators such as $\nabla \cdot \mathbf{F}$ and $\nabla \times \mathbf{F}$ can be expressed in the coordinates (σ, τ) by substituting the scale factors into the general formulae found in orthogonal coordinates.

Applications

The classic applications of bipolar coordinates are in solving partial differential equations, e.g., Laplace's equation or the Helmholtz equation, for which bipolar coordinates allow a separation of variables. A typical example would be the electric field surrounding two parallel cylindrical conductors.

Bispherical Coordinates

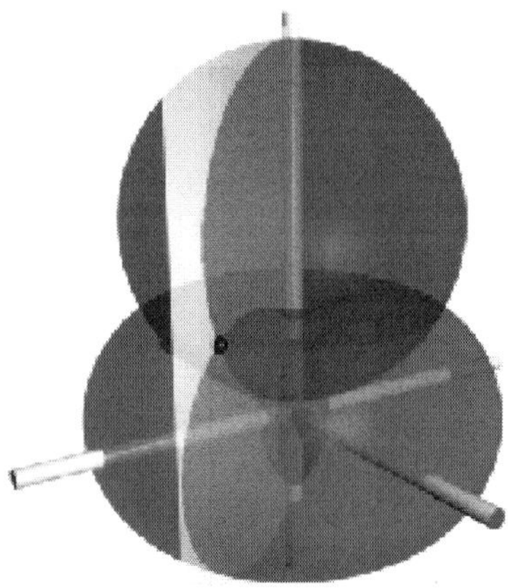

Illustration of bispherical coordinates, which are obtained by rotating a two-dimensional bipolar coordinate system about the axis joining its two foci. The foci are located at distance 1 from the vertical z-axis. The red self-interecting torus is the $\sigma = 45°$ isosurface, the blue sphere is the $\tau = 0.5$ isosurface, and the yellow half-plane is the $\varphi = 60°$ isosurface. The green half-plane marks the x-z plane, from which φ is measured. The black point is located at the intersection of the red, blue and yellow isosurfaces, at Cartesian coordinates roughly (0.841, -1.456, 1.239).

Bispherical coordinates are a three-dimensional orthogonal coordinate system that results from rotating the two-dimensional bipolar coordinate system about the axis that connects the two foci. Thus, the two foci F_1 and F_2 in bipolar coordinates remain points (on the z-axis, the axis of rotation) in the bispherical coordinate system.

Hyperbolic Coordinates

In mathematics, hyperbolic coordinates are a method of locating points in Quadrant I of the Cartesian plane

$$\{(x, y) : x > 0, y > 0\} = Q$$

Hyperbolic coordinates take values in the hyperbolic plane defined as:

$$HP = \{(u, v) : u \in \square, v > 0\}$$

These coordinates in HP are useful for studying logarithmic comparisons of direct proportion in Q and measuring deviations from direct proportion.

For (x, y) in Q take

$$u = -\frac{1}{2}\ln\left(\frac{y}{x}\right)$$

and

$$v = \sqrt{xy}$$

Sometimes the parametre u is called hyperbolic angle and v the geometric mean.

The inverse mapping is

$$x = ve^{u}, \quad y = ve^{-u}$$

This is a continuous mapping, but not an analytic function.

Quadrant Model of Hyperbolic Geometry

The correspondence

$$Q \leftrightarrow HP$$

affords the hyperbolic geometry structure to Q that is erected on HP by hyperbolic motions. The *hyperbolic lines* in Q are rays from the origin or petal-shaped curves leaving and re-entering the origin. The left-right shift in HP corresponds to a squeeze mapping applied to Q. Note that hyperbolas in Q do *not* represent geodesics in this model.

If one only considers the Euclidean topology of the plane and the topology inherited by Q, then the lines bounding Q seem close to Q. Insight from the metric space *HP* shows that the open set Q has only the origin as boundary when viewed as the quadrant model of the hyperbolic plane. Indeed, consider rays from the origin in Q, and their images, vertical rays from the boundary R of *HP*. Any point in *HP* is an infinite distance from the point p at the foot of the perpendicular to R, but a sequence of points on this perpendicular may tend in the direction of p. The corresponding sequence in Q tends along a ray towards the origin. The old Euclidean boundary of Q is irrelevant to the quadrant model.

Applications in Physical Science

Physical unit relations like:

- $V = I\ R$: Ohm's law
- $P = V\ I$: Electrical power
- $P\ V = k\ T$: Ideal gas law
- $f\ \lambda = c$: Sine waves

All suggest looking carefully at the quadrant. For example, in thermodynamics the isothermal process explicitly follows the hyperbolic path and work can be interpreted as a hyperbolic angle change. Similarly, an isobaric process may trace a hyperbola in the quadrant of absolute temperature and gas density.

Statistical Applications

- Comparative study of population density in the quadrant begins with selecting a reference nation, region, or urban area whose population and area are taken as the point (1,1).
- Analysis of the elected representation of regions in a representative democracy begins with selection of a standard for comparison: a particular represented group, whose magnitude and slate magnitude (of representatives) stands at (1,1) in the quadrant.

Economic Applications

There are many natural applications of hyperbolic coordinates in economics:

Analysis of currency exchange rate fluctuation:

The unit currency sets $x = 1$. The price currency corresponds to y. For

$$0 < y < 1$$

we find $u > 0$, a positive hyperbolic angle. For a *fluctuation* take a new price

$$0 < z < y$$

Then the change in u is:

$$\Delta u = \frac{1}{2}\log\left(\frac{y}{z}\right)$$

Quantifying exchange rate fluctuation through hyperbolic angle provides an objective, symmetric, and consistent measure. The quantity Δu is the length of the left-right shift in the hyperbolic motion view of the currency fluctuation.

- Analysis of inflation or deflation of prices of a basket of consumer goods.
- Quantification of change in marketshare in duopoly.
- Corporate stock splits versus stock buy-back.

History

While the geometric mean is an ancient concept, the hyperbolic angle is contemporary with the development of logarithm, the latter part of the seventeenth century. Gregoire de Saint-Vincent, Marin Mersenne, and Alphonse Antonio de Sarasa evaluated the quadrature of the hyperbola as a function having properties now familiar for the logarithm. The exponential function, the hyperbolic sine, and the hyperbolic cosine followed. As complex function theory referred to infinite series the circular functions sine and cosine seemed to absorb the hyperbolic sine and cosine as depending on an imaginary variable. In the nineteenth century biquaternions came into use and exposed the alternative complex plane called split-complex numbers where the hyperbolic angle is raised to a level equal to the classical angle. In English literature biquaternions were used to model spacetime and show its symmetries. There the hyperbolic angle parametre came to be called rapidity. For relativists, examining the quadrant as the possible future between oppositely directed photons, the geometric mean parametre is temporal.

In relativity the focus is on the 3-dimensional hypersurface in the future of spacetime where various velocities arrive after a given proper time. Scott Walter explains that in November 1907 Herman Minkowski alluded to a well-known three-dimensional hyperbolic geometry while speaking to the Göttingen Mathematical Society, but not to a four-dimensional one. In tribute to Wolfgang Rindler, the author of the standard introductory university-level textbook on relativity, hyperbolic coordinates of spacetime are called Rindler coordinates.

Identity Line

In a 2-dimensional Cartesian coordinate system, with x representing the abscissa and y the ordinate, the identity line is the $y = x$ line. The line, sometimes called the 1:1 line, has a slope of 1. When the abscissa and ordinate are on the same scale, the identity line forms a 45° angle with the abscissa, and is thus also, informally, called the 45° line. The line is often used as a reference in a 2-dimensional scatter plot comparing two sets of data expected to be identical under ideal conditions. When the corresponding data points from the two data sets are equal to each other, the corresponding scatters fall exactly on the identity line.

Chapter

4

Line Coordinates

In geometry, line coordinates are used to specify the position of a line just as point coordinates (or simply coordinates) are used to specify the position of a point.

Lines in the Plane

There are several possible ways to specify the position of a line in the plane. A simple way is by the pair (m, b) where the equation of the line is $y = mx + b$. Here m is the slope and b is the x-intercept. This system specifies coordinates for all lines that are not vertical. However, it is more common and simpler algebraically to use coordinates (l, m) where the equation of the line is $lx + my + 1 = 0$. This system specifies coordinates for all lines except those that pass through the origin. The geometrical interpretations of l and m are the negative reciprocals of the x and y-intercept respectively.

The exclusion of lines passing through the origin can be resolved by using a system of three coordinates (l, m, n) to specify the line in which the equation, $lx + my + n = 0$. Here l and m may not both be 0. In this equation, only the ratios between l, m and n are significant, in other words if the coordinates are multiplied by a non-zero scalar then line represented remains the same. So (l, m, n) is a system of homogeneous coordinates for the line.

If points in the plane are represented by homogeneous coordinates (x, y, z), the equation of the line is $lx + my + nz = 0$. In this context, l, m and n may not all be 0. In particular, $(0, 0, 1)$ represents the line

$z = 0$, which is the line at infinity in the projective plane. The coordinates (0, 1, 0) and (1, 0, 0) represent the x and y-axes respectively.

Tangential Equations

Just as $f(x, y) = 0$ can represent a curve as a subset of the points in the plane, the equation $\varphi(l, m) = 0$ represents a subset of the lines on the plain. The set of lines on the plane may, in an abstract sense, be thought of as the set of points in a projective plane, the dual of the original plane. The equation $\varphi(l, m) = 0$ then represents a curve in the dual plane.

For a curve $f(x, y) = 0$ in the plane, the tangents to the curve form a curve in the dual space called the dual curve.

If $\varphi(l, m) = 0$ is the equation of the dual curve, then it is called the tangential equation, for the original curve. A given equation $\varphi(l, m) = 0$ represents a curve in the original plane determined as the envelope of the lines that satisfy this equation. Similarly, if $\varphi(l, m, n)$ is a homogeneous function then $\varphi(l, m, n) = 0$ represents a curve in the dual space given in homogeneous coordinates, and may be called the homogeneous tangential equation of the enveloped curve. Tangential equations are useful in the study of curves defined as envelopes, just as Cartesian equations are useful in the study of curves defined as loci.

Tangential Equation of a Point

A linear equation in line coordinates has the form $al + bm + c = 0$, where a, b and c are constants. Suppose (l, m) is a line that satisfies this equation. If c is not 0 then $lx + my + 1 = 0$, where $x = a/c$ and $y = b/c$, so every line satisfying the original equation passes though the point (x, y). Conversely, any line through (x, y) satisfies the original equation, so $al + bm + c = 0$ is the equation of set of lines through (x, y). For a given point (x, y), the equation of the set of lines though it is $lx + my + 1 = 0$, so this may be defined as the tangential equation of the point. Similarly, for a point (x, y, z) given in homogeneous coordinates, then the equation of the point in homogeneous tangential coordinates is $(lx, my, nz) = 0$.

Formulas

The intersection of the lines (l_1, m_1) and (l_2, m_2) is the solution to the linear equations

$$l_1x + m_1y + 1 = 0$$

$$l_2x + m_2y + 1 = 0.$$

By Cramer's rule, the solution is

$$x = \frac{m_1 - m_2}{l_1 m_2 - l_2 m_1}, y = -\frac{l_1 - l_2}{l_1 m_2 - l_2 m_1}.$$

The lines (l_1, m_1), (l_2, m_2), and (l_3, m_3) are concurrent when the determinant

$$\begin{vmatrix} l_1 & m_1 & 1 \\ l_2 & m_2 & 1 \\ l_3 & m_3 & 1 \end{vmatrix} = 0.$$

For homogeneous coordinates, the intersection of the lines (l_1, m_1, n_1) and (l_2, m_2, n_2) is

$$(m_1 n_2 - m_2 n_1, l_2 n_1 - l_1 n_2, l_1 m_2 - l_2 m_1).$$

The lines (l_1, m_1, n_1), (l_2, m_2, n_2) and (l_3, m_3, n_3) are concurrent when the determinant

$$\begin{vmatrix} l_1 & m_1 & n_1 \\ l_2 & m_2 & n_2 \\ l_3 & m_3 & n_3 \end{vmatrix} = 0.$$

Dually, the coordinates of the line containing (x_1, y_1, z_1) and (x_2, y_2, z_2) are

$$(y_1 z_2 - y_2 z_1, x_2 z_1 - x_1 z_2, x_1 y_2 - x_2 y_1).$$

Lines in Three-dimensional Space

For two given points in the plane, (x_1, y_1, z_1) and (x_2, y_2, z_2), the three determinants

$$x_1 y_2 - x_2 y_1, x_1 z_2 - x_2 z_1, y_1 z_2 - y_2 z_1$$

determine the line containing them. Similarly, for two points in three-dimensional space (x_1, y_1, z_1, w_1) and (x_2, y_2, z_2, w_2), the line containing them is determined by the six determinants

$$x_1 y_2 - x_2 y_1, x_1 z_2 - x_1 z_2, y_1 z_2 - y_2 z_1, x_1 w_2 -$$
$$x_2 w_1, y_1 w_2 - y_2 w_1, z_1 w_2 - z_2 w_1.$$

This is the basis for a system of homogeneous line coordinates in three-dimensional space called *Plücker coordinates*. Six numbers in a set of coordinates only represent a line when they satisfy an additional equation. This system maps the space of lines in three-dimensional space to a projective space of dimension five, but with the additional requirement the space of lines is a manifold of dimension four.

More generally, the lines in n-dimensional projective space are determined by a system of $n(n-1)/2$ homogeneous coordinates that satisfy a set of $(n-2)(n-3)/2$ conditions, resulting in a manifold of dimension $2(n-1)$.

Log-Polar Coordinates

Log-polar coordinates (or logarithmic polar coordinates) is a coordinate system in two dimensions, where a point is identified by two numbers, one for the logarithm of the distance to a certain point, and one for an angle. Log-polar coordinates are closely connected to polar coordinates, which are usually used to describe domains in the plane with some sort of rotational symmetry. In areas like harmonic and complex analysis, the log-polar coordinates are more canonical than polar coordinates.

Definition and Coordinate Transformations

Log-polar coordinates in the plane consist of a pair of real numbers (ρ,θ), where ρ is the logarithm of the distance to a given point (the origin) and θ is the angle between a line of reference (the x-axis) and the line through the origin and the point. The angular coordinate is the same as for polar coordinates, while the radial coordinate is transformed according to the rule

$$r = e^{\rho}.$$

where r is the distance to the origin. The formulas for transformation from Cartesian coordinates to log-polar coordinates are given by

$$\begin{cases} \rho = \log\sqrt{x^2+y^2}, \\ \theta = \arctan y/x \text{ if } x > 0. \end{cases}$$

and the formulas for transformation from log-polar to Cartesian coordinates are

$$\begin{cases} x = e^{\rho}\cos\theta, \\ y = e^{\rho}\sin\theta. \end{cases}$$

By using complex numbers $(x, y) = x + iy$, the latter transformation can be written as

$$x + iy = e^{\rho + i\theta}$$

i.e. the complex exponential function. From this follows that basic equations in harmonic and complex analysis will have the same simple form as in Cartesian coordinates. This is not the case for polar coordinates.

Some Important Equations in Log-polar Coordinates

Laplace's Equation

Laplace's equation in two dimensions is given by

$$\frac{\partial^2 u}{\partial x^2}+\frac{\partial^2 u}{\partial y^2}=0$$

in Cartesian coordinates. Writing the same equation in polar coordinates gives the more complicated equation

$$r\frac{\partial}{\partial r}\left(r\frac{\partial u}{\partial r}\right)+\frac{\partial^2 u}{\partial \theta^2}=0$$

or equivalently

$$\left(r\frac{\partial}{\partial r}\right)^2 u+\frac{\partial^2 u}{\partial \theta^2}=0$$

However, from the relation $r = e^{\rho}$ it follows that $r\frac{\partial}{\partial r}=\frac{\partial}{\partial \rho}$ so Laplace's equation in log-polar coordinates,

$$\frac{\partial^2 u}{\partial \rho^2}+\frac{\partial^2 u}{\partial \theta^2}=0$$

has the same simple expression as in Cartesian coordinates. This is true for all coordinate systems where the transformation to Cartesian coordinates is given by a conformal mapping.

Thus, when considering Laplace's equation for a part of the plane with rotational symmetry, e.g. a circular disk, log-polar coordinates is the natural choice.

Dirichlet-to-Neumann Function

The latter coordinate system is for instance suitable for dealing with Dirichlet-to-Neumann functions. If the discrete coordinate system is interpreted as an undirected graph in the unit disc, it can be considered as a model for an electrical network.

To every line segment in the graph is associated a conductance given by a function γ. The electrical network will then serve as a discrete model for the Dirichlet problem in the unit disc, where the Laplace equation takes the form of Kirchhoff's law. On the nodes on the boundary of the circle, an electrical potential (Dirichlet data) is defined, which induces an electrical current (Neumann data) through the boundary nodes. The linear function Λ_γ mapping Dirichlet data to

Neumann data is called a Dirichlet-to-Neumann function, and depends on the topology and conductance of the network.

In the case with the continuous disc, it follows that if the conductance is homogeneous, let's say $\gamma = 1$ everywhere, then the Dirichlet-to-Neumann function satisfies the following equation

$$\Lambda_\gamma^2 + \frac{\partial^2}{\partial\theta^2} = 0$$

In order to get a good discrete model of the Dirichlet problem, it would be useful to find a graph in the unit disc whose (discrete) Dirichlet-to-Neumann function has the same property. Even though polar coordinates don't give us any answer, this is exactly what the spiral network given by log-polar coordinates provides us with.

Image Analysis

Already at the end of the 1970s, applications for the discrete spiral coordinate system were given in image analysis. To represent an image in this coordinate system rather than in Cartesian coordinates, gives computational advantages when rotating or zooming in an image. Also, the photo receptors in the retina in the human eye are distributed in a way that has big similarities with the spiral coordinate system. It can also be found in the Mandelbrot fractal.

Log-polar coordinates can also be used to construct fast methods for the Radon transform and its inverse.

Orthogonal Coordinates

In mathematics, orthogonal coordinates are defined as a set of d coordinates q = $(q^1, q^2, ..., q^d)$ in which the coordinate surfaces all meet at right angles (note: superscripts are indices, not exponents). A coordinate surface for a particular coordinate q^k is the curve, surface, or hypersurface on which q_k is a constant. For example, the three-dimensional Cartesian coordinates (x, y, z) is an orthogonal coordinate system, since its coordinate surfaces x = constant, y = constant, and z = constant are planes that meet at right angles to one another, i.e., are perpendicular. Orthogonal coordinates are a special but extremely common case of curvilinear coordinates. While vector operations and physical laws are normally easiest to derive in Cartesian coordinates, non-Cartesian orthogonal coordinates are often used instead for the solution of various problems, especially boundary value problems, such as those arising in field theories of quantum mechanics, fluid flow, electrodynamics and the diffusion of chemical species or heat.

The chief advantage of non-Cartesian coordinates is that they can be chosen to match the symmetry of the problem. For example, the pressure wave due to an explosion far from the ground (or other barriers) depends on 3D space in Cartesian coordinates, however the pressure predominantly moves away from the centre, so that in spherical coordinates the problem becomes very nearly one dimensional (since the pressure wave dominantly depends only on time and the distance from the centre). Another example is (slow) fluid in a straight circular pipe: in Cartesian coordinates, one has to solve a (difficult) two dimensional boundary value problem involving a partial differential equation, but in cylindrical coordinates the problem becomes one dimensional with an ordinary differential equation instead of a partial differential equation.

The reason to prefer orthogonal coordinates instead of general curvilinear coordinates is simplicity: many complications arise when coordinates are not orthogonal. For example, in orthogonal coordinates many problems may be solved by separation of variables. Separation of variables is a mathematical technique that converts a complex d-dimensional problem into d one-dimensional problems that can be solved in terms of known functions. Many equations can be reduced to Laplace's equation or the Helmholtz equation. Laplace's equation is separable in 13 orthogonal coordinate systems, and the Helmholtz equation is separable in 11 orthogonal coordinate systems.

Orthogonal coordinates never have off-diagonal terms in their metric tensor. In other words, the infinitesimal squared distance ds^2 can always be written as a scaled sum of the squared infinitesimal coordinate displacements

$$ds^2 = \sum_{k=1}^{d} \left(h_k\, dq^k\right)^2$$

where d is the dimension and the scaling functions (or scale factors)

$$h_k(\mathbf{q}) \overset{\text{def}}{=} \sqrt{g_{kk}(\mathbf{q})} = |\mathbf{e}_k|$$

equal the square roots of the diagonal components of the metric tensor, or the lengths of the local basis vectors e_k described below. These scaling functions h_i are used to calculate differential operators in the new coordinates, e.g., the gradient, the Laplacian, the divergence and the curl.

A simple method for generating orthogonal coordinates systems in two dimensions is by a conformal mapping of a standard two-dimensional grid of Cartesian coordinates (x, y). A complex number $z = x + iy$ can be formed from the real coordinates x and y, where i represents the square root of -1. Any holomorphic function $w = f(z)$ with non-zero complex derivative will produce a conformal mapping; if the resulting complex number is written $w = u + iv$, then the curves of constant u and v intersect at right angles, just as the original lines of constant x and y did.

Orthogonal coordinates in three and higher dimensions can be generated from an orthogonal two-dimensional coordinate system, either by projecting it into a new dimension (*cylindrical coordinates*) or by rotating the two-dimensional system about one of its symmetry axes. However, there are other orthogonal coordinate systems in three dimensions that cannot be obtained by projecting or rotating a two-dimensional system, such as the ellipsoidal coordinates. More general orthogonal coordinates may be obtained by starting with some necessary coordinate surfaces and considering their orthogonal trajectories.

Basis Vectors

Covariant Vasis

In Cartesian coordinates, the basis vectors are fixed (constant). In the more general setting of curvilinear coordinates, a point in space is specified by the coordinates, and at every such point there is bound a set of basis vectors, which generally are not constant: this is the essence of curvilinear coordinates in general and is a very important concept. What distinguishes orthogonal coordinates is that, though the basis vectors vary, they are always orthogonal with respect to each other. In other words,

$$\mathbf{e}_i \cdot \mathbf{e}_j = 0 \quad \text{if} \quad i \neq j$$

These basis vectors are by definition the tangent vectors of the curves obtained by varying one coordinate, keeping the others fixed:

$$\mathbf{e}_i = \frac{\partial \mathbf{r}}{\partial q^i}$$

where r is some point and q^i is the coordinate for which the basis vector is extracted. In other words, a curve is obtained by fixing all but one coordinate; the unfixed coordinate is varied as in a parametric curve, and the derivative of the curve with respect to the parametre (the varying coordinate) is the basis vector for that coordinate.

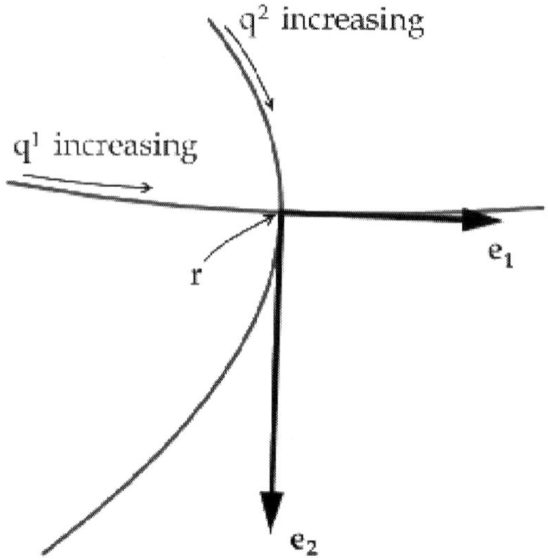

Figure: *Visualization of 2D orthogonal coordinates. Curves obtained by holding all but one coordinate constant are shown, along with basis vectors. Note that the basis vectors aren't of equal length: they need not be, they only need to be orthogonal.*

Note that the vectors are not necessarily of equal length. The normalised basis vectors are notated with a hat and obtained by dividing by the length:

$$\hat{\mathbf{e}}_i = \frac{\mathbf{e}_i}{|\mathbf{e}_i|}$$

A vector field may be specified by its components with respect to the basis vectors or the normalised basis vectors, one must be sure which case is dealt. Components in the normalised basis are most common in applications for clarity of the quantities (for example, one may want to deal with tangential velocity instead of tangential velocity times a scale factor); in derivations the normalised basis is less common since it is more complicated.

The useful functions known as scale factors (sometimes called Lamé coefficients, this should be avoided since some more well known coefficients in linear elasticity carry the same name) of the coordinates are simply the lengths of the basis vectors.

Contravariant Basis

The basis vectors shown above are covariant basis vectors (because they "co-vary" with vectors). In the case of orthogonal coordinates, the contravariant basis vectors are easy to find since they will be in the same direction as the covariant vectors but reciprocal length (for this reason, the two sets of basis vectors are said to be reciprocal with respect to each other):

$$\mathbf{e}^i = \frac{\hat{\mathbf{e}}_i}{h_i} = \frac{\mathbf{e}_i}{h_i^2}$$

this follows from the fact that, by definition, $\mathbf{e}_i \cdot \mathbf{e}_j = \delta_i^j$, using the Kronecker delta. Note that:

$$\hat{\mathbf{e}}_i = \frac{\mathbf{e}_i}{h_i} = h_i \mathbf{e}^i = \hat{\mathbf{e}}^i$$

We now face three different basis sets commonly used to describe vectors in orthogonal coordinates: the covariant basis e_i, the contravariant basis e^i, and the normalised basis κ^i. While a vector is an *objective quantity*, meaning its identity is independent of any coordinate system, the components of a vector depend on what basis the vector is represented in.

To avoid confusion, the components of the vector x with respect to the e_i basis are represented as x^i, while the components with respect to the e^i basis are represented as x_i:

$$\mathbf{x} = \sum x^i \mathbf{e}_i = \sum x_i \mathbf{e}^i$$

The position of the indices represent how the components are calculated (upper indices should not be confused with exponentiation). Note that the summation symbols Σ (capital Sigma) and the summation range, indicating summation over all basis vectors ($i = 1, 2, ..., d$), are often omitted. The components are related simply by:

$$h_i^2 x^i = x_i$$

There is no distinguishing widespread notation in use for vector components with respect to the normalised basis; in this article we'll use subscripts for vector components and note that the components are calculated in the normalised basis.

Vector Algebra

Vector addition and negation are done component-wise just as in Cartesian coordinates with no complication. Extra considerations may be necessary for other vector operations.

Note however, that all of these operations assume that two vectors in a vector field are bound to the same point (in other words, the tails of vectors coincide). Since basis vectors generally vary in orthogonal coordinates, if two vectors are added whose components are calculated at different points in space, the different basis vectors require consideration.

Dot Product

The dot product in Cartesian coordinates (Euclidean space with an orthonormal basis set) is simply the sum of the products of components. In orthogonal coordinates, the dot product of two vectors x and y takes this familiar form when the components of the vectors are calculated in the normalised basis:

$$\mathbf{x}\cdot\mathbf{y} = \sum x_i \hat{\mathbf{e}}_i \cdot \sum y_i \hat{\mathbf{e}}_i = \sum x_i y_i$$

This is an immediate consequence of the fact that the normalised basis at some point can form a Cartesian coordinate system: the basis set is orthonormal.

For components in the covariant or contraviant bases,

$$\mathbf{x}\cdot\mathbf{y} = \sum h_i^2 x^i y^i = \sum \frac{x_i y_i}{h_i^2} = \sum x^i y_i = \sum x_i y^i$$

This can be readily derived by writing out the vectors in component form, normalising the basis vectors, and taking the dot product. For example, in 2D:

$$\begin{aligned}\mathbf{x}\cdot\mathbf{y} &= \left(x^1\mathbf{e}_1 + x^2\mathbf{e}_2\right)\cdot\left(y_1\mathbf{e}^1 + y_2\mathbf{e}^2\right) \\ [10pt] &= \left(x^1 h_1 \hat{\mathbf{e}}_1 + x^2 h_2 \hat{\mathbf{e}}_2\right)\cdot\left(y_1 \frac{\hat{\mathbf{e}}^1}{h_1} + y_2 \frac{\hat{\mathbf{e}}^2}{h_2}\right) = x^1 y_1 + x^2 y_2\end{aligned}$$

where the fact that the normalised covariant and contravariant bases are equal has been used.

Cross Product

The cross product in 3D Cartesian coordinates is:

$$\mathbf{x}\times\mathbf{y} = (x_2 y_3 - x_3 y_2)\hat{\mathbf{e}}_1 + (x_3 y_1 - x_1 y_3)\hat{\mathbf{e}}_2 + (x_1 y_2 - x_2 y_1)\hat{\mathbf{e}}_3$$

The above formula then remains valid in orthogonal coordinates if the components are calculated in the normalised basis.

To construct the cross product in orthogonal coordinates with covariant or contravariant bases we again must simply normalise the basis vectors, for example:

$$\mathbf{x}\times\mathbf{y} = \sum x^i \mathbf{e}_i \times \sum y^i \mathbf{e}_i = \sum x^i h_i \hat{\mathbf{e}}_i \times \sum y^i h_i \hat{\mathbf{e}}_i$$

which, written expanded out,

$$\mathbf{x}\times\mathbf{y} = (x^2y^3 - x^3y^2)\frac{h_2h_3}{h_1}\mathbf{e}_1 + (x^3y^1 - x^1y^3)\frac{h_1h_3}{h_2}\mathbf{e}_2 + (x^1y^2 - x^2y^1)\frac{h_1h_2}{h_3}\mathbf{e}_3$$

Terse notation for the cross product, which simplifies generalisation to non-orthogonal coordinates and higher dimensions, is possible with the Levi-Civita tensor, which will have components other than zeros and ones if the scale factors are not all equal to one.

Vector Calculus

Differentiation

Looking at an infinitesimal displacement from some point, it's apparent that

$$d\mathbf{r} = \sum \frac{\partial \mathbf{r}}{\partial q^i} dq^i = \sum \mathbf{e}_i\, dq^i$$

By definition, the gradient of a function must satisfy (this definition remains true if f is any tensor)

$$df = \nabla f \cdot d\mathbf{r} \quad \Rightarrow \quad df = \nabla f \cdot \sum \mathbf{e}_i\, dq^i$$

It follows then that del operator must be:

$$\nabla = \sum \mathbf{e}^i \frac{\partial}{\partial q^i}$$

and this happens to remain true in general curvilinear coordinates. Quantities like the gradient and Laplacian follow through proper application of this operator.

Differential Operators in Three Dimensions

Since these operations are common in application, all vector components in this section are presented with respect to the normalised basis.

The gradient of a scalar equals

$$\nabla\phi = \frac{\hat{\mathbf{e}}_1}{h_1}\frac{\partial\phi}{\partial q^1} + \frac{\hat{\mathbf{e}}_2}{h_2}\frac{\partial\phi}{\partial q^2} + \frac{\hat{\mathbf{e}}_3}{h_3}\frac{\partial\phi}{\partial q^3}.$$

The Laplacian of a scalar equals

$$\nabla^2\phi = \frac{1}{h_1h_2h_3}\left[\frac{\partial}{\partial q^1}\left(\frac{h_2h_3}{h_1}\frac{\partial\phi}{\partial q^1}\right) + \frac{\partial}{\partial q^2}\left(\frac{h_3h_1}{h_2}\frac{\partial\phi}{\partial q^2}\right) + \frac{\partial}{\partial q^3}\left(\frac{h_1h_2}{h_3}\frac{\partial\phi}{\partial q^3}\right)\right].$$

The divergence of a vector equals

$$\nabla\cdot\mathbf{F} = \frac{1}{h_1h_2h_3}\left[\frac{\partial}{\partial q^1}\left(F_1h_2h_3\right)+\frac{\partial}{\partial q^2}\left(F_2h_3h_1\right)+\frac{\partial}{\partial q^3}\left(F_3h_1h_2\right)\right].$$

The curl equals

$$\begin{aligned}\nabla\times\mathbf{F} &= \frac{\hat{\mathbf{e}}_1}{h_2h_3}\left[\frac{\partial}{\partial q^2}\left(h_3F_3\right)-\frac{\partial}{\partial q^3}\left(h_2F_2\right)\right] \\ &+\frac{\hat{\mathbf{e}}_2}{h_3h_1}\left[\frac{\partial}{\partial q^3}\left(h_1F_1\right)-\frac{\partial}{\partial q^1}\left(h_3F_3\right)\right] \\ &+\frac{\hat{\mathbf{e}}_3}{h_1h_2}\left[\frac{\partial}{\partial q^1}\left(h_2F_2\right)-\frac{\partial}{\partial q^2}\left(h_1F_1\right)\right] \\ [10pt] &= \frac{1}{h_1h_2h_3}\begin{vmatrix} h_1\hat{\mathbf{e}}_1 & h_2\hat{\mathbf{e}}_2 & h_3\hat{\mathbf{e}}_3 \\ \frac{\partial}{\partial q^1} & \frac{\partial}{\partial q^2} & \frac{\partial}{\partial q^3} \\ h_1F_1 & h_2F_2 & h_3F_3 \end{vmatrix}.\end{aligned}$$

Parabolic Cylindrical Coordinates

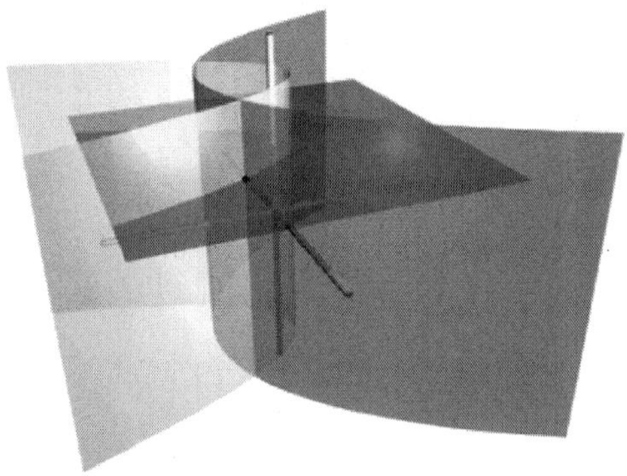

Figure: *Coordinate surfaces of parabolic cylindrical coordinates. The red parabolic cylinder corresponds to σ=2, whereas the yellow parabolic cylinder corresponds to τ=1. The blue plane corresponds to* z=2*. These surfaces intersect at the point P (shown as a black sphere), which has Cartesian coordinates roughly (2, -1.5, 2).*

In mathematics, parabolic cylindrical coordinates are a three-dimensional orthogonal coordinate system that results from projecting the two-dimensional parabolic coordinate system in the perpendicular *z*-direction. Hence, the coordinate surfaces are confocal parabolic cylinders. Parabolic cylindrical coordinates have found many applications, e.g., the potential theory of edges.

Basic Definition

The parabolic cylindrical coordinates (σ, τ, z) are defined in terms of the Cartesian coordinates *(x,y,z)* by:

$$x = \sigma\tau$$

$$y = \frac{1}{2}\left(\tau^2 - \sigma^2\right)$$

$$z = z$$

The surfaces of constant σ form confocal parabolic cylinders

$$2y = \frac{x^2}{\sigma^2} - \sigma^2$$

that open towards $+y$, whereas the surfaces of constant τ form confocal parabolic cylinders

$$2y = -\frac{x^2}{\tau^2} + \tau^2$$

that open in the opposite direction, i.e., towards $-y$. The foci of all these parabolic cylinders are located along the line defined by $x = y = 0$. The radius r has a simple formula as well

$$r = \sqrt{x^2 + y^2} = \frac{1}{2}\left(\sigma^2 + \tau^2\right)$$

that proves useful in solving the Hamilton-Jacobi equation in parabolic coordinates for the inverse-square central force problem of mechanics.

Scale Factors

The scale factors for the parabolic cylindrical coordinates σ and τ are:

$$h_\sigma = h_\tau = \sqrt{\sigma^2 + \tau^2}$$

$$h_z = 1$$

The infinitesimal element of volume is

$$dV = h_\sigma h_\tau h_z = \left(\sigma^2 + \tau^2\right) d\sigma d\tau dz$$

and the Laplacian equals

$$\nabla^2 \Phi = \frac{1}{\sigma^2 + \tau^2}\left(\frac{\partial^2 \Phi}{\partial \sigma^2} + \frac{\partial^2 \Phi}{\partial \tau^2}\right) + \frac{\partial^2 \Phi}{\partial z^2}$$

Other differential operators such as $\nabla \cdot \mathbf{F}$ and $\nabla \times \mathbf{F}$ can be expressed in the coordinates (σ, τ) by substituting the scale factors into the general formulae found in orthogonal coordinates.

Parabolic Cylinder Harmonics

Since all of the surfaces of constant σ, τ and z are conicoid, Laplace's equation is separable in parabolic cylindrical coordinates. Using the technique of the separation of variables, a separated solution to Laplace's equation may be written:

$$V = S(\sigma)T(\tau)Z(z)$$

and Laplace's equation, divided by V, is written:

$$\frac{1}{\sigma^2+\tau^2}\left[\frac{\ddot{S}}{S}+\frac{\ddot{T}}{T}\right]+\frac{\ddot{Z}}{Z}=0$$

Since the Z equation is separate from the rest, we may write

$$\frac{\ddot{Z}}{Z}=-m^2$$

where m is constant. $Z(z)$ has the solution:

$$Z_m(z) = A_1 e^{imz} + A_2 e^{-imz}$$

Substituting $-m^2$ for $\ddot{Z}/Z$, Laplace's equation may now be written:

$$\left[\frac{\ddot{S}}{S}+\frac{\ddot{T}}{T}\right]=m^2(\sigma^2+\tau^2)$$

We may now separate the S and T functions and introduce another constant n^2 to obtain:

$$\ddot{S}-(m^2\sigma^2+n^2)S=0$$

$$\ddot{T}-(m^2\tau^2-n^2)T=0$$

The solutions to these equations are the parabolic cylinder functions

$$S_{mn}(\sigma) = A_3\, y_1(n^2/2m, \sigma\sqrt{2m}) + A_4\, y_2(n^2/2m, \sigma\sqrt{2m})$$

$$T_{mn}(\tau) = A_5\, y_1(n^2/2m, i\tau\sqrt{2m}) + A_6\, y_2(n^2/2m, i\tau\sqrt{2m})$$

The parabolic cylinder harmonics for (m,n) are now the product of the solutions. The combination will reduce the number of constants and the general solution to Laplace's equation may be written:

$$V(\sigma,\tau,z) = \sum_{m,n} A_{mn} S_{mn} T_{mn} Z_m$$

Applications

The classic applications of parabolic cylindrical coordinates are in solving partial differential equations, e.g., Laplace's equation or the

Helmholtz equation, for which such coordinates allow a separation of variables. A typical example would be the electric field surrounding a flat semi-infinite conducting plate.

Two-centre Bipolar Coordinates

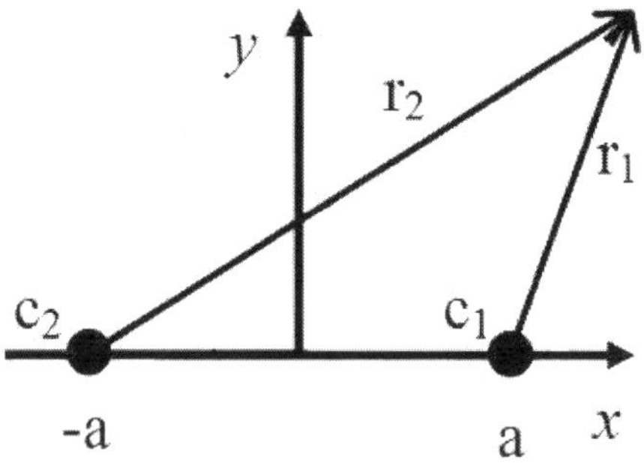

Figure: *Two-centre bipolar coordinates.*

In mathematics, two-centre bipolar coordinates is a coordinate system, based on two coordinates which give distances from two fixed centres, C_1 and C_2 This system is very useful in some scientific applications It should not be confused with so-called bipolar coordinates.

Cartesian coordinates

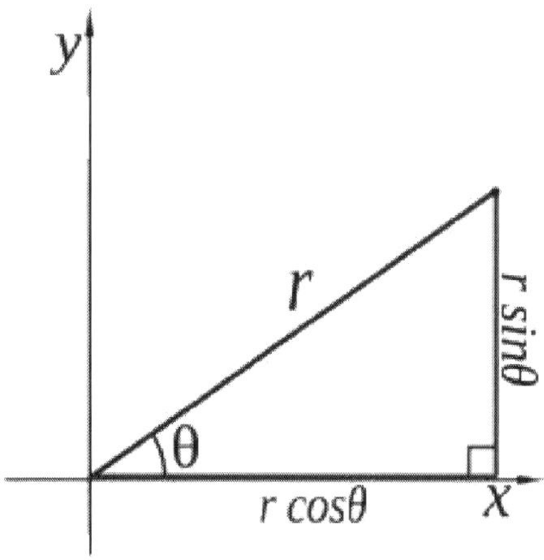

Figure: *Cartesian coordinates and polar coordinates.*

Transformation to Cartesian coordinates (x, y) from two-centre bipolar coordinates (r_1, r_2)

$$x = \frac{r_1^2 - r_2^2}{4a}$$

$$y = \pm\frac{1}{4a}\sqrt{16a^2 r_1^2 - (r_1^2 - r_2^2 + 4a^2)^2}$$

where the centres of this coordinate system are at (+a, 0) and (-a, 0).

Polar Coordinates

To polar coordinates from two-centre bipolar coordinates

$$r = \sqrt{\frac{r_1^2 + r_2^2 - 2a^2}{2}}$$

$$\theta = \arctan\left[\frac{\sqrt{8a^2(r_1^2 + r_2^2 - 2a^2) - (r_1^2 - r_2^2)^2}}{r_1^2 - r_2^2}\right]$$

Where $2a$ is the distance between the poles (coordinate system centres).

Darboux's Theorem

Darboux's theorem is a theorem in the mathematical field of differential geometry and more specifically differential forms, partially generalising the Frobenius integration theorem. It is a foundational result in several fields, the chief among them being symplectic geometry. The theorem is named after Jean Gaston Darboux who established it as the solution of the Pfaff problem. One of the many consequences of the theorem is that any two symplectic manifolds of the same dimension are locally symplectomorphic to one another. That is, every $2n$-dimensional symplectic manifold can be made to look locally like the linear symplectic space C^n with its canonical symplectic form. There is also an analogous consequence of the theorem as applied to contact geometry.

Statement and First Consequences

The precise statement is as follows. Suppose that θ is a differential 1-form on an n dimensional manifold, such that dθ has constant rank p. If

$$\theta \wedge (d\theta)^p = 0 \text{ everywhere,}$$

then there is a local system of coordinates $x_1,\ldots,x_{n-p}, y_1, \ldots, y_p$ in which

$$\theta = x_1 dy_1 + \ldots + x_p dy_p.$$

If, on the other hand,

$$\theta \wedge (d\theta)^p \neq 0 \text{ everywhere,}$$

then there is a local system of coordinates $x_1,...,x_{n-p}, y_1, ..., y_p$ in which

$$\theta = x_1 dy_1 + ... + x_p dy_p + dx_{p+1}.$$

In particular, suppose that ω is a symplectic 2-form on an $n = 2m$ dimensional manifold M. In a neighbourhood of each point p of M, by the Poincaré lemma, there is a 1-form θ with dθ = ω. Moreover, θ satisfies the first set of hypotheses in Darboux's theorem, and so locally there is a coordinate chart U near p in which

$$\theta = x_1 dy_1 + ... + x_m dy_m.$$

Taking an exterior derivative now shows

$$\omega = d\theta = dx_1 \wedge dy_1 + ... + dx_m \wedge dy_m.$$

The chart U is said to be a Darboux chart around p. The manifold M can be covered by such charts.

To state this differently, identify R^{2m} with C^m by letting $z_j = x_j + i\, y_j$. If $\varphi : U \to C^n$ is a Darboux chart, then ω is the pullback of the standard symplectic form ω_0 on C^n:

$$\omega = \phi^* \omega_0.$$

Comparison with Riemannian Geometry

This result implies that there are no local invariants in symplectic geometry: a Darboux basis can always be taken, valid near any given point.

This is in marked contrast to the situation in Riemannian geometry where the curvature is a local invariant, an obstruction to the metric being locally a sum of squares of coordinate differentials.

The difference is that Darboux's theorem states that ω can be made to take the standard form in an *entire neighbourhood* around p. In Riemannian geometry, the metric can always be made to take the standard form *at* any given point, but not always in a neighbourhood around that point.

Isothermal Coordinates

In mathematics, specifically in differential geometry, isothermal coordinates on a Riemannian manifold are local coordinates where the metric is conformal to the Euclidean metric. This means that in isothermal coordinates, the Riemannian metric locally has the form

$$g = e^{\varphi}(dx_1^2 + \cdots + dx_n^2),$$

where φ is a smooth function.

Isothermal coordinates on surfaces were first introduced by Gauss. Korn and Lichtenstein proved that isothermal coordinates exist around any point on a two dimensional Riemannian manifold. On higher dimensional Riemannian manifolds a necessary and sufficient condition for their local existence is the vanishing of the Weyl tensor and of the Cotton tensor.

Isothermal Coordinates on Surfaces

The first result on the existence of isothermal coordinates was due to Korn (1916) and Lichtenstein (1916). Later accounts were given by Morrey (1938), Ahlfors (1955), Bers (1952) and Chern (1955). A particularly simple account using the Hodge star operator is given in DeTurck & Kazdan (1981).

Beltrami Equation

The existence of isothermal coordinates can be proved by applying known existence theorems for the Beltrami equation, which rely on L^p estimates for singular integral operators of Calderon and Zygmund. A simpler approach to the Beltrami equation has been given more recently by the late Adrien Douady.

If the Riemannian metric is given locally as

$$ds^2 = Edx^2 + 2Fdxdy + Gdy^2,$$

then in the complex coordinate $z = x + iy$, it takes the form

$$ds^2 = \lambda\,|dz + \mu d\bar{z}|^2,$$

where λ and μ are smooth with $\lambda > 0$ and $|\mu| < 1$. In fact

$$\lambda = \frac{1}{4}(E + G + 2\sqrt{EG - F^2}),\ \mu = (E - G + 2iF)/\lambda.$$

In isothermal coordinates (u, v) the metric should take the form

$$ds^2 = \rho(du^2 + dv^2)$$

with $\rho > 0$ smooth. The complex coordinate $w = u + i\,v$ satisfies

$$\rho|\,dw\,|^2 = \rho\,|\,w_z\,|^2|dz + \frac{w_{\bar{z}}}{w_z}d\bar{z}\,|^2,$$

so that the coordinates (u, v) will be isothermal if the Beltrami equation

$$\frac{\partial w}{\partial \bar{z}} = \mu \frac{\partial w}{\partial z}$$

has a diffeomorphic solution. Such a solution has been proved to exist in any neighbourhood where $||\mu||_{-} < 1$.

Gaussian Curvature

In the isothermal coordinates (u, v), the Gaussian curvature takes the simpler form

$$K = -\frac{1}{2} e^{-\varphi} \left(\frac{\partial^2 \varphi}{\partial u^2} + \frac{\partial^2 \varphi}{\partial v^2} \right),$$

where $\rho = e^{\varphi}$.

Normal Coordinates

In differential geometry, normal coordinates at a point p in a differentiable manifold equipped with a symmetric affine connection are a local coordinate system in a neighbourhood of p obtained by applying the exponential map to the tangent space at p. In a normal coordinate system, the Christoffel symbols of the connection vanish at the point p, thus often simplifying local calculations. In normal coordinates associated to the Levi-Civita connection of a Riemannian manifold, one can additionally arrange that the metric tensor is the Kronecker delta at the point p, and that the first partial derivatives of the metric at p vanish.

A basic result of differential geometry states that normal coordinates at a point always exist on a manifold with a symmetric affine connection. In such coordinates the covariant derivative reduces to a partial derivative (at p only), and the geodesics through p are locally linear functions of t (the affine parametre). This idea was implemented in a fundamental way by Albert Einstein in the general theory of relativity: the equivalence principle uses normal coordinates via inertial frames. Normal coordinates always exist for the Levi-Civita connection of a Riemannian or Pseudo-Riemannian manifold. By contrast, there is no way to define normal coordinates for Finsler manifolds (Busemann 1955).

Geodesic Normal Coordinates

Geodesic normal coordinates are local coordinates on a manifold with an affine connection afforded by the exponential map

$$\exp_p : T_p M \supset V \to M$$

and an isomorphism

$$E : \mathbb{R}^n \to T_p M$$

given by any basis of the tangent space at the fixed basepoint $p - M$. If the additional structure of a Riemannian metric is imposed, then the basis defined by E may be required in addition to be orthonormal, and the resulting coordinate system is then known as a Riemannian normal coordinate system.

Normal coordinates exist on a normal neighbourhood of a point p in M.

A normal neighbourhood U is a subset of M such that there is a proper neighbourhood V of the origin in the tangent space T_pM and $\exp_p$ acts as a diffeomorphism between U and V. Now let U be a normal neighbourhood of p in M then the chart is given by:

$$\varphi := E^{-1} \circ \exp_p^{-1} : U \to \mathbb{R}^n$$

The isomorphism E can be any isomorphism between both vectorspaces, so there are as many charts as different orthonormal bases exist in the domain of E.

Polar Coordinates

On a Riemannian manifold, a normal coordinate system at p facilitates the introduction of a system of spherical coordinates, known as polar coordinates. These are the coordinates on M obtained by introducing the standard spherical coordinate system on the Euclidean space T_pM. That is, one introduces on T_pM the standard spherical coordinate system (r, φ) where $r \geq 0$ is the radial parametre and $\varphi = (\varphi_1,\ldots,\varphi_{n-1})$ is a parameterization of the $(n-1)$-sphere. Composition of (r, φ) with the inverse of the exponential map at p is a polar coordinate system. Polar coordinates provide a number of fundamental tools in Riemannian geometry. The radial coordinate is the most significant: geometrically it represents the geodesic distance to p of nearby points. Gauss's lemma asserts that the gradient of r is simply the partial derivative $\partial/\partial r$. That is,

$$\langle df, dr \rangle = \frac{\partial f}{\partial r}$$

for any smooth function f. As a result, the metric in polar coordinates assumes a block diagonal form

$$g = \begin{bmatrix} 1 & 0 & \cdots 0 \\ 0 & & \\ \vdots & & g_{\phi\phi}(r,\phi) \\ 0 & & \end{bmatrix}.$$

Chapter

5

Derivations of the Quadratic Formula

By Completing the Square

The quadratic formula can be derived by the method of completing the square, so as to make use of the algebraic identity:

$$x^2 + 2xh + h^2 = (x+h)^2.$$

Dividing the quadratic equation

$$ax^2 + bx + c = 0$$

by a (which is allowed because a is non-zero), gives:

$$x^2 + \frac{b}{a}x + \frac{c}{a} = 0,$$

or

$$x^2 + \frac{b}{a}x = -\frac{c}{a}.$$

The quadratic equation is now in a form to which the method of completing the square can be applied. To "complete the square" is to add a constant to both sides of the equation such that the left hand side becomes a complete square:

$$x^2 + \frac{b}{a}x + \left(\frac{1}{2}\frac{b}{a}\right)^2 = -\frac{c}{a} + \left(\frac{1}{2}\frac{b}{a}\right)^2,$$

which produces

$$\left(x+\frac{b}{2a}\right)^2=-\frac{c}{a}+\frac{b^2}{4a^2}.$$

The right side can be written as a single fraction, with common denominator $4a^2$. This gives

$$\left(x+\frac{b}{2a}\right)^2=\frac{b^2-4ac}{4a^2}.$$

Taking the square root of both sides yields

$$x+\frac{b}{2a}=\pm\frac{\sqrt{b^2-4ac}}{2a}.$$

Isolating x, gives

$$x=-\frac{b}{2a}\pm\frac{\sqrt{b^2-4ac}}{2a}=\frac{-b\pm\sqrt{b^2-4ac}}{2a}.$$

Other Methods of Root Calculation

Alternative Quadratic Formula

In some situations it is preferable to express the roots in an alternative form.

$$x=\frac{2c}{-b\mp\sqrt{b^2-4ac}}=\frac{2c}{-b\mp\sqrt{\Delta}}.$$

This alternative requires c to be nonzero; for, if c is zero, the formula correctly gives zero as one root, but fails to give any second, non-zero root. Instead, one of the two choices for –" produces the indeterminate form 0/0, which is undefined. However, the alternative form works when a is zero (giving the unique solution as one root and division by zero again for the other), which the normal form does not (instead producing division by zero both times).

The roots are the same regardless of which expression we use; the alternative form is merely an algebraic variation of the common form:

$$\frac{-b+\sqrt{b^2-4ac}}{2a}=\frac{-b+\sqrt{b^2-4ac}}{2a}\cdot\frac{-b-\sqrt{b^2-4ac}}{-b-\sqrt{b^2-4ac}}$$

$$=\frac{b^2-(b^2-4ac)}{2a\left(-b-\sqrt{b^2-4ac}\right)}$$

$$= \frac{4ac}{2a\left(-b-\sqrt{b^2-4ac}\right)}$$

$$= \frac{2c}{-b-\sqrt{b^2-4ac}}.$$

The alternative formula can reduce loss of precision in the numerical evaluation of the roots, which may be a problem if one of the roots is much smaller than the other in absolute magnitude. In this case, b is very close to $\pm\sqrt{x}$, and the subtraction in the numerator causes loss of significance.

A mixed approach avoids both all cancellation problems (only numbers of the same sign are added), and the problem of c being zero:

$$x_1 = \frac{-b-\operatorname{sgn}(b)\sqrt{b^2-4ac}}{2a},$$

$$x_2 = \frac{2c}{-b-\operatorname{sgn}(b)\sqrt{b^2-4ac}} = \frac{c}{ax_1}.$$

Here sgn denotes the sign function.

Floating Point Implementation

A careful floating point computer implementation differs a little from both forms to produce a robust result. Assuming the discriminant, $b^2 - 4ac$, is positive and b is nonzero, the code will be something like the following:

$$q = -\tfrac{1}{2}\left(b+\operatorname{sgn}(b)\sqrt{b^2-4ac}\right)$$

$$x_1 = q/a$$

$$x_2 = c/q$$

Here sgn(b) is the sign function, where sgn(b) is 1 if b is positive and −1 if b is negative; its use ensures that the quantities added are of the same sign, avoiding *catastrophic cancellation*. The computation of x_2 uses the fact that the product of the roots is c/a. Note that while the above formulation avoids catastrophic cancellation between b and $\sqrt{b^2-4ac}$ there remains a form of cancellation between the terms b^2 and $-4ac$ of the discriminant, which can still lead to loss of up to half of correct significant figures. The discriminant b^2-4ac needs to be computed in arithmetic of twice the precision of the result to avoid this (e.g. quad precision if the final result is to be accurate to full double precision).

Generalisation of Quadratic Equation

The formula and its derivation remain correct if the coefficients a, b and c are complex numbers, or more generally members of any field whose characteristic is not 2. (In a field of characteristic 2, the element $2a$ is zero and it is impossible to divide by it.)

The symbol

$$\pm\sqrt{b^2 - 4ac}$$

in the formula should be understood as "either of the two elements whose square is $b^2 - 4ac$, if such elements exist". In some fields, some elements have no square roots and some have two; only zero has just one square root, except in fields of characteristic 2. Note that even if a field does not contain a square root of some number, there is always a quadratic extension field which does, so the quadratic formula will always make sense as a formula in that extension field.

Characteristic 2

In a field of characteristic 2, the quadratic formula, which relies on 2 being a unit, does not hold. Consider the monic quadratic polynomial

$$x^2 + bx + c$$

over a field of characteristic 2. If $b = 0$, then the solution reduces to extracting a square root, so the solution is

$$x = \sqrt{c}$$

and note that there is only one root since

$$-\sqrt{c} = -\sqrt{c} + 2\sqrt{c} = \sqrt{c}.$$

In summary,

$$x^2 + c = (x + \sqrt{c})^2.$$

In the case that b ‘Š 0, there are two distinct roots, but if the polynomial is irreducible, they cannot be expressed in terms of square roots of numbers in the coefficient field. Instead, define the 2-root $R(c)$ of c to be a root of the polynomial $x^2 + x + c$, an element of the splitting field of that polynomial. One verifies that $R(c) + 1$ is also a root. In terms of the 2-root operation, the two roots of the (non-monic) quadratic $ax^2 + bx + c$ are

$$\frac{b}{a}R\left(\frac{ac}{b^2}\right)$$

and

$$\frac{b}{a}\left(R\left(\frac{ac}{b^2}\right)+1\right).$$

For example, let a denote a multiplicative generator of the group of units of F_4, the Galois field of order four (thus a and $a + 1$ are roots of $x^2 + x + 1$ over F_4). Because $(a + 1)^2 = a$, $a + 1$ is the unique solution of the quadratic equation $x^2 + a = 0$. On the other hand, the polynomial $x^2 + ax + 1$ is irreducible over F_4, but it splits over F_{16}, where it has the two roots ab and $ab + a$, where b is a root of $x^2 + x + a$ in F_{16}.

Trigonometric Functions

In mathematics, the trigonometric functions (also called circular functions) are functions of an angle. They are used to relate the angles of a triangle to the lengths of the sides of a triangle. Trigonometric functions are important in the study of triangles and modelling periodic phenomena, among many other applications.

The most familiar trigonometric functions are the sine, cosine, and tangent. In the context of the standard unit circle with radius 1, where a triangle is formed by a ray originating at the origin and making some angle with the x-axis, the sine of the angle gives the length of the y-component (rise) of the triangle, the cosine gives the length of the x-component (run), and the tangent function gives the slope (y-component divided by the x-component).

More precise definitions are detailed below. Trigonometric functions are commonly defined as ratios of two sides of a right triangle containing the angle, and can equivalently be defined as the lengths of various line segments from a unit circle. More modern definitions express them as infinite series or as solutions of certain differential equations, allowing their extension to arbitrary positive and negative values and even to complex numbers.

Trigonometric functions have a wide range of uses including computing unknown lengths and angles in triangles (often right triangles). In this use, trigonometric functions are used, for instance, in navigation, engineering, and physics. A common use in elementary physics is resolving a vector into Cartesian coordinates. The sine and cosine functions are also commonly used to model periodic function phenomena such as sound and light waves, the position and velocity of harmonic oscillators, sunlight intensity and day length, and average temperature variations through the year.

In modern usage, there are six basic trigonometric functions, tabulated here with equations that relate them to one another. Especially with the last four, these relations are often taken as the *definitions* of those functions, but one can define them equally well geometrically, or by other means, and then derive these relations.

Right-angled Triangle Definitions

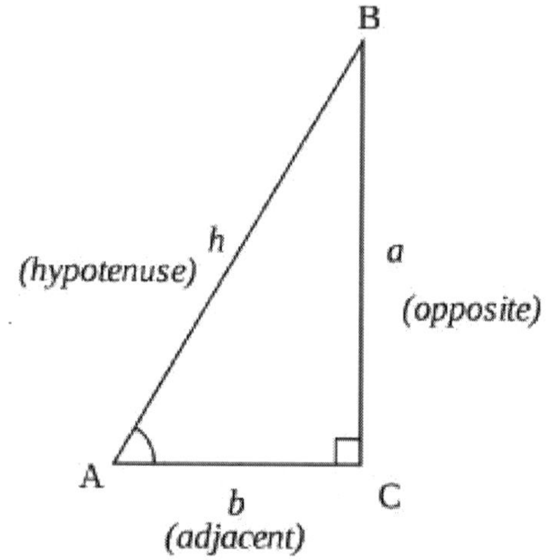

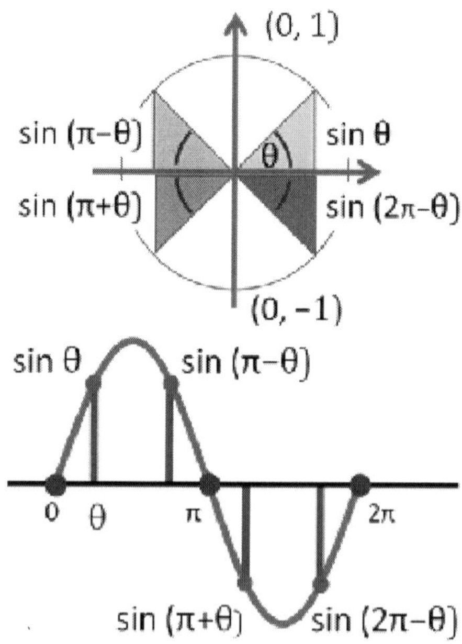

Figure: *(Top): Trigonometric function sinè for selected angles* θ, $\pi - \theta$, $\pi + \theta$, *and* $2\pi - \theta$ *in the four quadrants. (Bottom) Graph of sine function versus angle. Angles from the top panel are identified.*

The notion that there should be some standard correspondence between the lengths of the sides of a triangle and the angles of the triangle comes as soon as one recognises that similar triangles maintain the same ratios between their sides. That is, for any similar triangle the ratio of the hypotenuse (for example) and another of the sides remains the same. If the hypotenuse is twice as long, so are the sides. It is these ratios that the trigonometric functions express. To define the trigonometric functions for the angle A, start with any right triangle that contains the angle A. The three sides of the triangle are named as follows:

- The *hypotenuse* is the side opposite the right angle, in this case side h. The hypotenuse is always the longest side of a right-angled triangle.
- The *opposite side* is the side opposite to the angle we are interested in (angle A), in this case side a.
- The *adjacent side* is the side having both the angles of interest (angle A and right-angle C), in this case side b.

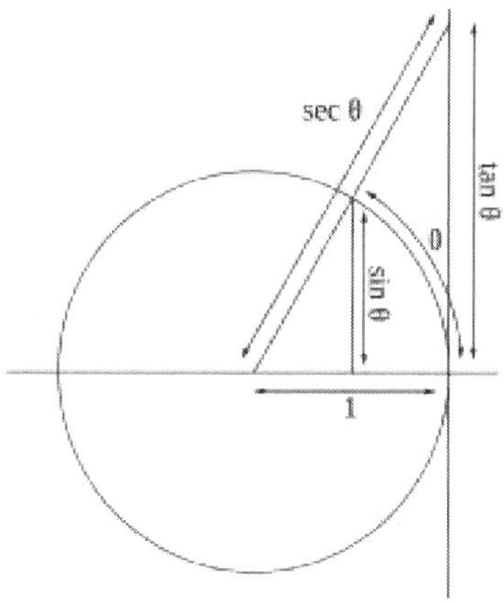

Figure: *The trigonometric functions are summarised in the following table and described in more detail below. The angle θ is the angle between the hypotenuse and the adjacent line – the angle at A in the accompanying diagram. The sine, tangent, and secant functions of an angle constructed geometrically in terms of a unit circle. The number q is the length of the curve; thus angles are being measured in radians. The secant and tangent functions rely on a fixed vertical line and the sine function on a moving vertical line. ("Fixed" in this context means not moving as θ changes; "moving" means depending on θ.) Thus, as θ goes from 0 up to a right angle, sin θ goes from 0 to 1, tan θ goes from 0 to ∞, and sec θ goes from 1 to ∞.*

In ordinary Euclidean geometry, according to the triangle postulate, the inside angles of every triangle total 180° (π radians). Therefore, in a right-angled triangle, the two non-right angles total 90° ($\pi/2$ radians), so each of these angles must be in the range of (0°,90°) as expressed in interval notation. The following definitions apply to angles in this 0° – 90° range. They can be extended to the full set of real arguments by using the unit circle, or by requiring certain symmetries and that they be periodic functions. For example, the figure shows sin θ for angles θ, $\pi - \theta$, $\pi + \theta$, and $2\pi - \theta$ depicted on the unit circle (top) and as a graph (bottom). The value of the sine repeats itself apart from sign in all four quadrants, and if the range of θ is extended to additional rotations, this behaviour repeats periodically with a period 2π.

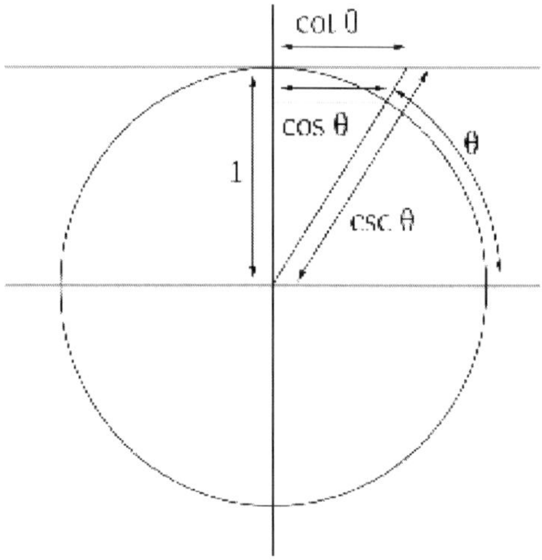

Figure: *The cosine, cotangent, and cosecant functions of an angle θ constructed geometrically in terms of a unit circle. The functions whose names have the prefix* co- *use horizontal lines where the others use vertical lines.*

Sine, Cosine and Tangent

The sine of an angle is the ratio of the length of the opposite side to the length of the hypotenuse. (The word comes from the Latin *sinus* for gulf or bay, since, given a unit circle, it is the side of the triangle on which the angle *opens*). In our case

$$\sin A = \frac{\text{opposite}}{\text{hypotenuse}} = \frac{a}{h}.$$

Note that this ratio does not depend on size of the particular right triangle chosen, as long as it contains the angle A, since all such triangles are similar.

The cosine of an angle is the ratio of the length of the adjacent side to the length of the hypotenuse: so called because it is the sine of the complementary or co-angle. In our case

$$\cos A = \frac{\text{adjacent}}{\text{hypotenuse}} = \frac{b}{h}.$$

The tangent of an angle is the ratio of the length of the opposite side to the length of the adjacent side: so called because it can be represented as a line segment tangent to the circle, that is the line that touches the circle, from Latin *linea tangens* or touching line (cf. *tangere*, to touch). In our case

$$\tan A = \frac{\text{opposite}}{\text{adjacent}} = \frac{a}{b}.$$

The acronyms "SOHCAHTOA" ("Soak-a-toe", "Sock-a-toa", "Soak-a-toa") and "OHSAHCOAT" are commonly used mnemonics for these ratios.

Reciprocal Functions

The remaining three functions are best defined using the above three functions. The cosecant csc(A), or cosec(A), is the reciprocal of sin(A), i.e. the ratio of the length of the hypotenuse to the length of the opposite side:

$$\csc A = \frac{1}{\sin A} = \frac{\text{hypotenuse}}{\text{opposite}} = \frac{h}{a}.$$

The secant sec(A) is the reciprocal of cos(A), i.e. the ratio of the length of the hypotenuse to the length of the adjacent side:

$$\sec A = \frac{1}{\cos A} = \frac{\text{hypotenuse}}{\text{adjacent}} = \frac{h}{b}.$$

It is so called because it represents the line that *cuts* the circle (from Latin: *secare*, to cut). The cotangent cot(A) is the reciprocal of tan(A), i.e. the ratio of the length of the adjacent side to the length of the opposite side:

$$\cot A = \frac{1}{\tan A} = \frac{\text{adjacent}}{\text{opposite}} = \frac{b}{a}.$$

Slope Definitions

Equivalent to the right-triangle definitions, the trigonometric functions can also be defined in terms of the *rise*, *run*, and *slope* of a line segment relative to horizontal. The slope is commonly taught as "rise over run" or D $_{\text{run}}$. The three main trigonometric functions are commonly taught in the order sine, cosine, tangent. With a line segment length of 1 (as in a unit circle), the following mnemonic devices show the correspondence of definitions:

1. "Sine is first, rise is first" meaning that Sine takes the angle of the line segment and tells its vertical rise when the length of the line is 1.
2. "Cosine is second, run is second" meaning that Cosine takes the angle of the line segment and tells its horizontal run when the length of the line is 1.
3. "Tangent combines the rise and run" meaning that Tangent takes the angle of the line segment and tells its slope; or alternatively, tells the vertical rise when the line segment's horizontal run is 1.

This shows the main use of tangent and arctangent: converting between the two ways of telling the slant of a line, *i.e.,* angles and slopes. (Note that the arctangent or "inverse tangent" is not to be confused with the *cotangent,* which is cosine divided by sine.)

While the length of the line segment makes no difference for the slope (the slope does not depend on the length of the slanted line), it does affect rise and run. To adjust and find the actual rise and run when the line does not have a length of 1, just multiply the sine and cosine by the line length. For instance, if the line segment has length 5, the run at an angle of 7° is 5 cos(7°)

Series Definitions

Trigonometric functions are analytic functions. Using only geometry and properties of limits, it can be shown that the derivative of sine is cosine and the derivative of cosine is the negative of sine. (Here, and generally in calculus, all angles are measured in radians.) One can then use the theory of Taylor series to show that the following identities hold for all real numbers x:

$$\begin{aligned}\sin x &= x-\frac{x^3}{3!}+\frac{x^5}{5!}-\frac{x^7}{7!}+\cdots \\ &= \sum_{n=0}^{\infty}\frac{(-1)^n x^{2n+1}}{(2n+1)!}, \\ \cos x &= 1-\frac{x^2}{2!}+\frac{x^4}{4!}-\frac{x^6}{6!}+\cdots \\ &= \sum_{n=0}^{\infty}\frac{(-1)^n x^{2n}}{(2n)!}.\end{aligned}$$

These identities are sometimes taken as the *definitions* of the sine and cosine function.

They are often used as the starting point in a rigorous treatment of trigonometric functions and their applications (e.g., in Fourier series), since the theory of infinite series can be developed, independent of any geometric considerations, from the foundations of the real number system. The differentiability and continuity of these functions are then established from the series definitions alone.

Combining these two series gives Euler's formula: cos x + i sin $x = e^{ix}$.

Other series can be found. For the following trigonometric functions:

U_n is the nth up/down number,

B_n is the nth Bernoulli number, and

E_n (below) is the nth Euler number.

Tangent

$$\begin{aligned}\tan x &= \sum_{n=0}^{\infty} \frac{U_{2n+1} x^{2n+1}}{(2n+1)!} \\ &= \sum_{n=1}^{\infty} \frac{(-1)^{n-1} 2^{2n} (2^{2n}-1) B_{2n} x^{2n-1}}{(2n)!} \\ &= x + \frac{1}{3}x^3 + \frac{2}{15}x^5 + \frac{17}{315}x^7 + \cdots, \qquad \text{for } |x| < \frac{\pi}{2}.\end{aligned}$$

When this series for the tangent function is expressed in a form in which the denominators are the corresponding factorials, the numerators, called the "tangent numbers", have a combinatorial interpretation: they enumerate alternating permutations of finite sets of odd cardinality.

Cosecant

$$\begin{aligned}\csc x &= \sum_{n=0}^{\infty} \frac{(-1)^{n+1} 2(2^{2n-1}-1) B_{2n} x^{2n-1}}{(2n)!} \\ &= x^{-1} + \frac{1}{6}x + \frac{7}{360}x^3 + \frac{31}{15120}x^5 + \cdots, \qquad \text{for } 0 < |x| < \pi.\end{aligned}$$

Secant

$$\begin{aligned}\sec x &= \sum_{n=0}^{\infty} \frac{U_{2n} x^{2n}}{(2n)!} = \sum_{n=0}^{\infty} \frac{(-1)^n E_{2n} x^{2n}}{(2n)!} \\ &= 1 + \frac{1}{2}x^2 + \frac{5}{24}x^4 + \frac{61}{720}x^6 + \cdots, \qquad \text{for } |x| < \frac{\pi}{2}.\end{aligned}$$

When this series for the secant function is expressed in a form in which the denominators are the corresponding factorials, the numerators, called the "secant numbers", have a combinatorial interpretation: they enumerate alternating permutations of finite sets of even cardinality.

Cotangent

$$\cot x = \sum_{n=0}^{\infty} \frac{(-1)^n 2^{2n} B_{2n} x^{2n-1}}{(2n)!}$$

$$= x^{-1} - \frac{1}{3}x - \frac{1}{45}x^3 - \frac{2}{945}x^5 - \cdots, \qquad \text{for } 0 < |x| < \pi.$$

From a theorem in complex analysis, there is a unique analytic continuation of this real function to the domain of complex numbers. They have the same Taylor series, and so the trigonometric functions are defined on the complex numbers using the Taylor series above.

There is a series representation as partial fraction expansion where just translated reciprocal functions are summed up, such that the poles of the cotangent function and the reciprocal functions match:

$$\pi \cdot \cot(\pi x) = \lim_{N \to \infty} \sum_{n=-N}^{N} \frac{1}{x+n}.$$

This identity can be proven with the Herglotz trick. By combining the $-n$-th with the n-th term, it can be expressed as an absolutely convergent series:

$$\pi \cdot \cot(\pi x) = \frac{1}{x} + \sum_{n=1}^{\infty} \frac{2x}{x^2 - n^2}.$$

Computation

The computation of trigonometric functions is a complicated subject, which can today be avoided by most people because of the widespread availability of computers and scientific calculators that provide built-in trigonometric functions for any angle.

This section, however, describes details of their computation in three important contexts: the historical use of trigonometric tables, the modern techniques used by computers, and a few "important" angles where simple exact values are easily found.

The first step in computing any trigonometric function is range reduction—reducing the given angle to a "reduced angle" inside a small range of angles, say 0 to $\pi/2$, using the periodicity and symmetries of the trigonometric functions.

Prior to computers, people typically evaluated trigonometric functions by interpolating from a detailed table of their values, calculated to many significant figures. Such tables have been available for as long as trigonometric functions have been described, and were typically generated by repeated application of the half-angle and angle-addition identities starting from a known value (such as sin $(\pi/2) = 1$).

Modern computers use a variety of techniques. One common method, especially on higher-end processors with floating point units, is to combine a polynomial or rational approximation (such as Chebyshev approximation, best uniform approximation, and Padé approximation, and typically for higher or variable precisions, Taylor and Laurent series) with range reduction and a table lookup—they first look up the closest angle in a small table, and then use the polynomial to compute the correction. Devices that lack hardware multipliers often use an algorithm called CORDIC (as well as related techniques), which uses only addition, subtraction, bitshift, and table lookup. These methods are commonly implemented in hardware floating-point units for performance reasons. For very high precision calculations, when series expansion convergence becomes too slow, trigonometric functions can be approximated by the arithmetic-geometric mean, which itself approximates the trigonometric function by the (complex) elliptic integral. Finally, for some simple angles, the values can be easily computed by hand using the Pythagorean theorem, as in the following examples. For example, the sine, cosine and tangent of any integer multiple of $\pi/60$ radians (3°) can be found exactly by hand.

Consider a right triangle where the two other angles are equal, and therefore are both $\pi/4$ radians (45°). Then the length of side b and the length of side a are equal; we can choose $a = b = 1$. The values of sine, cosine and tangent of an angle of $\pi/4$ radians (45°) can then be found using the Pythagorean theorem:

$$c = \sqrt{a^2 + b^2} = \sqrt{2}.$$

Therefore:

$$\sin\frac{\pi}{4} = \sin 45^\circ = \cos\frac{\pi}{4} = \cos 45^\circ = \frac{1}{\sqrt{2}},$$

$$\tan\frac{\pi}{4} = \tan 45^\circ = \frac{\sin\frac{\pi}{4}}{\cos\frac{\pi}{4}} = \frac{1}{\sqrt{2}}.\frac{\sqrt{2}}{1} = \frac{\sqrt{2}}{\sqrt{2}} = 1.$$

To determine the trigonometric functions for angles of π/3 radians (60 degrees) and π/6 radians (30 degrees), we start with an equilateral triangle of side length 1. All its angles are π/3 radians (60 degrees). By dividing it into two, we obtain a right triangle with π/6 radians (30 degrees) and π/3 radians (60 degrees) angles. For this triangle, the shortest side = 1/2, the next largest side =(−3)/2 and the hypotenuse = 1. This yields:

$$\sin\frac{\pi}{6} = \sin 30^\circ = \cos\frac{\pi}{3} = \cos 60^\circ = \frac{1}{2},$$

$$\cos\frac{\pi}{6} = \cos 30^\circ = \sin\frac{\pi}{3} = \sin 60^\circ = \frac{\sqrt{3}}{2},$$

$$\tan\frac{\pi}{6} = \tan 30^\circ = \cot\frac{\pi}{3} = \cot 60^\circ = \frac{1}{\sqrt{3}}.$$

Properties and Applications

The trigonometric functions, as the name suggests, are of crucial importance in trigonometry, mainly because of the following two results.

Law of Sines

The law of sines states that for an arbitrary triangle with sides a, b, and c and angles opposite those sides A, B and C:

$$\frac{\sin A}{a} = \frac{\sin B}{b} = \frac{\sin C}{c},$$

or, equivalently,

$$\frac{a}{\sin A} = \frac{b}{\sin B} = \frac{c}{\sin C} = 2R,$$

where R is the triangle's circumradius.

It can be proven by dividing the triangle into two right ones and using the above definition of sine. The law of sines is useful for computing the lengths of the unknown sides in a triangle if two angles and one side are known. This is a common situation occurring in *triangulation*, a technique to determine unknown distances by measuring two angles and an accessible enclosed distance.

Law of Cosines

The law of cosines (also known as the cosine formula or cosine rule) is an extension of the Pythagorean theorem:

$$c^2 = a^2 + b^2 - 2ab\cos C,$$

or equivalently,

$$\cos C = \frac{a^2 + b^2 - c^2}{2ab}.$$

In this formula the angle at C is opposite to the side c. This theorem can be proven by dividing the triangle into two right ones and using the Pythagorean theorem. The law of cosines can be used to determine a side of a triangle if two sides and the angle between them are known. It can also be used to find the cosines of an angle (and consequently the angles themselves) if the lengths of all the sides are known.

Law of Tangents

The following all form the law of tangents

$$\frac{\tan\frac{A-B}{2}}{\tan\frac{A+B}{2}} = \frac{a-b}{a+b}$$

$$[6pt]\frac{\tan\frac{A-C}{2}}{\tan\frac{A+C}{2}} = \frac{a-c}{a+c}$$

$$[6pt]\frac{\tan\frac{B-C}{2}}{\tan\frac{B+C}{2}} = \frac{b-c}{b+c}$$

The explanation of the formulae in words would be cumbersome, but the patterns of sums and differences; for the lengths and corresponding opposite angles, are apparent in the theorem.

Law of Cotangents

If

$$\zeta = \sqrt{\frac{1}{s}(s-a)(s-b)(s-c)}$$

(the radius of the inscribed circle for the triangle) and

$$s = \frac{a+b+c}{2}$$

(the semi-perimetre for the triangle), then the following all form the law of cotangents

$$\cot\frac{A}{2} = \frac{s-a}{\zeta}$$

$$\cot\frac{B}{2}=\frac{s-b}{\zeta}$$

$$\cot\frac{C}{2}=\frac{s-c}{\zeta}$$

It follows that

$$\frac{\cot(A/2)}{s-a}=\frac{\cot(B/2)}{s-b}=\frac{\cot(C/2)}{s-c}.$$

In words the theorem is: the cotangent of a half-angle equals the ratio of the semi-perimetre minus the opposite side to the said angle, to the inradius for the triangle.

History

While the early study of trigonometry can be traced to antiquity, the trigonometric functions as they are in use today were developed in the medieval period. The chord function was discovered by Hipparchus of Nicaea (180–125 BC) and Ptolemy of Roman Egypt (90–165 AD).

The functions sine and cosine can be traced to the *jyâ* and *koti-jyâ* functions used in Gupta period Indian astronomy (*Aryabhatiya, Surya Siddhanta*), via translation from Sanskrit to Arabic and then from Arabic to Latin.

All six trigonometric functions in current use were known in Islamic mathematics by the 9th century, as was the law of sines, used in solving triangles. al-Khwârizmî produced tables of sines, cosines and tangents. They were studied by authors including Omar Khayyam, Bhâskara II, Nasir al-Din al-Tusi, Jamshîd al-Kâshî (14th century), Ulugh Beg (14th century), Regiomontanus (1464), Rheticus, and Rheticus' student Valentinus Otho

Madhava of Sangamagrama (c. 1400) made early strides in the analysis of trigonometric functions in terms of infinite series.

The first published use of the abbreviations 'sin', 'cos', and 'tan' is by the 16th century French mathematician Albert Girard.

In a paper published in 1682, Leibniz proved that sin x is not an algebraic function of x.

Leonhard Euler's *Introductio in analysin infinitorum* (1748) was mostly responsible for establishing the analytic treatment of trigonometric functions in Europe, also defining them as infinite series and presenting "Euler's formula", as well as the near-modern abbreviations *sin., cos., tang., cot., sec.,* and *cosec.*

A few functions were common historically, but are now seldom used, such as the chord (crd(θ) = 2 sin(θ/2)), the versine (versin(θ) = 1 − cos(θ) = 2 $\sin^2$(θ/2)) (which appeared in the earliest tables), the haversine (haversin(θ) = versin(θ) / 2 = $\sin^2$(θ/2)), the exsecant (exsec(θ) = sec(θ) − 1) and the excosecant (excsc(θ) = exsec(π/2 − θ) = csc(θ) − 1). Many more relations between these functions are listed in the article about trigonometric identities.

Etymologically, the word *sine* derives from the Sanskrit word for half the chord, *jya-ardha*, abbreviated to *jiva*. This was transliterated in Arabic as *jiba*, written *jb*, vowels not being written in Arabic. Next, this transliteration was mis-translated in the 12th century into Latin as *sinus*, under the mistaken impression that *jb* stood for the word *jaib*, which means "bosom" or "bay" or "fold" in Arabic, as does *sinus* in Latin. Finally, English usage converted the Latin word *sinus* to *sine*. The word *tangent* comes from Latin *tangens* meaning "touching", since the line *touches* the circle of unit radius, whereas *secant* stems from Latin *secans* — "cutting" — since the line *cuts* the circle.

Conic Section

In mathematics, a conic section (or just conic) is a curve obtained as the intersection of a cone (more precisely, a right circular conical surface) with a plane. In analytic geometry, a conic may be defined as a plane algebraic curve of degree 2. There are a number of other geometric definitions possible. One of the most useful, in that it involves only the plane, is that a conic consists of those points whose distances to some point, called a *focus*, and some line, called a *directrix*, are in a fixed ratio, called the *eccentricity*.

Traditionally, the three types of conic section are the hyperbola, the parabola, and the ellipse. The circle is a special case of the ellipse, and is of sufficient interest in its own right that it is sometimes called the fourth type of conic section.

The type of a conic corresponds to its eccentricity, those with eccentricity less than 1 being ellipses, those with eccentricity equal to 1 being parabolas, and those with eccentricity greater than 1 being hyperbolas. In the focus-directrix definition of a conic the circle is a limiting case with eccentricity 0. In modern geometry certain degenerate cases, such as the union of two lines, are included as conics as well.

The conic sections were named and studied at least since 200 BC, when Apollonius of Perga undertook a systematic study of their properties.

History

Menaechmus

It is believed that the first definition of a conic section is due to Menaechmus (died 320 BC). This work does not survive, however, and is only known through secondary accounts. The definition used at that time differs from the one commonly used today in that it requires the plane cutting the cone to be perpendicular to one of the lines that generate the cone as a surface of revolution (a generatrix). Thus the shape of the conic is determined by the angle formed at the vertex of the cone (between two opposite generatrices): If the angle is acute then the conic is an ellipse; if the angle is right then the conic is a parabola; and if the angle is obtuse then the conic is a hyperbola. Note that the circle cannot be defined this way and was not considered a conic at this time. Euclid (fl. 300 BC) is said to have written four books on conics but these were lost as well. Archimedes (died c. 212 BC) is known to have studied conics, having determined the area bounded by a parabola and an ellipse. The only part of this work to survive is a book on the solids of revolution of conics.

Apollonius of Perga

The greatest progress in the study of conics by the ancient Greeks is due to Apollonius of Perga (died c.190 BC)), whose eight volume *Conic Sections* summarised the existing knowledge at the time and greatly extended it. Apollonius's major innovation was to characterise a conic using properties within the plane and intrinsic to the curve; this greatly simplified analysis. With this tool, it was now possible to show that any plane cutting the cone, regardless of its angle, will produce a conic according to the earlier definition, leading to the definition commonly used today. Pappus of Alexandria (died c. 350 CE) is credited with discovering importance of the concept of a focus of a conic, and the discovery of the related concept of a directrix.

Al-Kuhi

An instrument for drawing conic sections was first described in 1000 CE by the Islamic mathematician Al-Kuhi.

Omar Khayyám

Apollonius's work was translated into Arabic (the technical language of the time) and much of his work only survives through the Arabic version. Persians found applications to the theory; the most notable of these was the Persian mathematician and poet Omar Khayyám who used conic sections to solve algebraic equations.

Europe

Johannes Kepler extended the theory of conics through the "principle of continuity", a precursor to the concept of limits. Girard Desargues and Blaise Pascal developed a theory of conics using an early form of projective geometry and this helped to provide impetus for the study of this new field. In particular, Pascal discovered a theorem known as the hexagrammum mysticum from which many other properties of conics can be deduced. Meanwhile, René Descartes applied his newly discovered Analytic geometry to the study of conics. This had the effect of reducing the geometrical problems of conics to problems in algebra.

Properties

Just as two (distinct) points determine a line, five points determine a conic. Formally, given any five points in the plane in general linear position, meaning no three collinear, there is a unique conic passing through them, which will be non-degenerate; this is true over both the affine plane and projective plane. Indeed, given any five points there is a conic passing through them, but if three of the points are collinear the conic will be degenerate (reducible, because it contains a line), and may not be unique.

Irreducible conic sections are always "smooth". This is important for many applications, such as aerodynamics, where a smooth surface is required to ensure laminar flow and to prevent turbulence.

Intersection at Infinity

An algebro-geometrically intrinsic form of this classification is by the intersection of the conic with the line at infinity, which gives further insight into their geometry:

- *Ellipses intersect the line at infinity in 0 points:* rather, in 0 real points, but in 2 complex points, which are conjugate;
- *Parabolas intersect the line at infinity in 1 double point, corresponding to the axis:* they are tangent to the line at infinity, and close at infinity, as distended ellipses;
- *Hyperbolas intersect the line at infinity in 2 points, corresponding to the asymptotes:* hyperbolas pass through infinity, with a twist. Going to infinity along one branch passes through the point at infinity corresponding to the asymptote, then re-emerges on the other branch at the other side but with the inside of the hyperbola (the direction of curvature) on the other side – left vs. right (corresponding to the non-orientability of

the real projective plane) – and then passing through the other point at infinity returns to the first branch. Hyperbolas can thus be seen as ellipses that have been pulled through infinity and re-emerged on the other side, flipped.

Degenerate Cases

There are five degenerate cases: three in which the plane passes through apex of the cone, and three that arise when the cone itself degenerates to a cylinder (a doubled line can occur in both cases).

When the plane passes through the apex, the resulting conic is always degenerate, and is either: a point (when the angle between the plane and the axis of the cone is larger than tangential); a straight line (when the plane is tangential to the surface of the cone); or a pair of intersecting lines (when the angle is smaller than the tangential).

These correspond respectively to degeneration of an ellipse, parabola, and a hyperbola, which are characterised in the same way by angle. The straight line is more precisely a *double* line (a line with multiplicity 2) because the plane is tangent to the cone, and thus the intersection should be counted twice.

Where the cone is a cylinder, i.e. with the vertex at infinity, cylindric sections are obtained; this corresponds to the apex being at infinity.

Cylindrical sections are ellipses (or circles), unless the plane is vertical (which corresponds to passing through the apex at infinity), in which case three degenerate cases occur: two parallel lines, known as a ribbon (corresponding to an ellipse with one axis infinite and the other axis real and non-zero, the distance between the lines), a double line (an ellipse with one infinite axis and one axis zero), and no intersection (an ellipse with one infinite axis and the other axis imaginary).

Cartesian Coordinates

In the Cartesian coordinate system, the graph of a quadratic equation in two variables is always a conic section – though it may be degenerate, and all conic sections arise in this way. The equation will be of the form

$$Ax^2 + Bxy + Cy^2 + Dx + Ey + F = 0 \text{ with } A, B, C \text{ not all zero.}$$

As scaling all six constants yields the same locus of zeros, one can consider conics as points in the five-dimensional projective space $\mathbf{P}^5$.

Matrix Representation of Conic Sections

In mathematics, the matrix representation of conic sections is one way of studying a conic section, its axis, vertices, foci, tangents, and the relative position of a given point.

We can also study conic sections whose axes aren't parallel to our coordinate system.

Conic sections have the form of a second-degree polynomial:

$$Q \overset{\text{def}}{=} Ax^2 + Bxy + Cy^2 + Dx + Ey + F = 0.$$

That can be written as:

$$\mathbf{x}^T A_Q \mathbf{x} = 0$$

Where x is the homogeneous coordinate vector:

$$\begin{pmatrix} x \\ y \\ 1 \end{pmatrix}$$

And A_Q a matrix:

$$A_Q = \begin{pmatrix} A & B/2 & D/2 \\ B/2 & C & E/2 \\ D/2 & E/2 & F \end{pmatrix}.$$

Classification

Regular and degenerated conic sections can be distinguished based on the determinant of A_Q.

If $\det A_Q = 0$, the conic is degenerate.

If Q is not degenerate, we can see what type of conic section it is by computing the minor $\det A_{33}$ (that is, the determinant of the submatrix resulting from removing the last row and the last column of A_Q):

$$A_{33} = \begin{bmatrix} A & B/2 \\ B/2 & C \end{bmatrix}.$$

- If and only if $\det A_{33} < 0$, it is a hyperbola.
- If and only if $\det A_{33} = 0$, it is a parabola.
- If and only if $\det A_{33} > 0$, it is an ellipse.

In the case of an ellipse, we can make a further distinction between an ellipse and a circle by comparing the last two diagonal elements corresponding to x^2 and y^2.

- If $A = C$ and $B = 0$, it is a circle. Moreover, in the case of a non-degenerate ellipse (with $\det A_{33} > 0$ and $\det A_Q \neq 0$), we have a real ellipse if $C \cdot \det A_Q < 0$ but an imaginary ellipse if $C \cdot \det A_Q > 0$. An example of the latter is $x^2 + y^2 + 10 = 0$, which has no real-valued solutions.

  If the conic section is degenerate ($\det A_Q = 0$), $\det A_{33}$ still allows us to distinguish its form:

- If and only if $\det A_{33} < 0$, it is two intersecting lines.
- If and only if $\det A_{33} = 0$, it is two parallel straight lines. These lines are distinct and real if $E^2 - 4CF > 0$, coincident if $E^2 - 4CF = 0$, and distinct and imaginary if $E^2 - 4CF < 0$.
- If and only if $\det A_{33} > 0$, it is a single point.

Centre

In the centre of the conic, the gradient of the quadratic form Q vanishes, so: $\nabla Q = [\frac{\partial Q}{\partial x}, \frac{\partial Q}{\partial y}] = [0,0]$.

We can calculate the centre by taking the first two rows of the associated matrix A_Q, multiplying each by $(x, y, 1)^T$, setting both inner products equal to 0, and solving the system.

$$S \overset{\text{def}}{=} \begin{cases} a_{11} + a_{12}x + a_{13}y = 0 \\ a_{21} + a_{22}x + a_{23}y = 0 \end{cases} \overset{\text{def}}{=} \begin{cases} D/2 + Ax + (B/2)y = 0 \\ E/2 + (B/2)x + Cy = 0 \end{cases}$$

This becomes

$$\begin{pmatrix} x_c \\ y_c \end{pmatrix} = \begin{pmatrix} A & B/2 \\ B/2 & C \end{pmatrix}^{-1} \begin{pmatrix} -D/2 \\ -E/2 \end{pmatrix} = \begin{pmatrix} (BE - 2CD)/(4AC - B^2) \\ (DB - 2AE)/(4AC - B^2) \end{pmatrix}$$

Note that in the case of a parabola, defined by $(4AC\text{-}B^2) = 0$, there is no centre since the above denominators become zero.

Axes

The major and minor axes are two lines determined by the centre of the conic as a point and eigenvectors of the associated matrix as vectors of direction.

$$a_{1,2} \stackrel{\text{def}}{=} \begin{cases} S(x_0, y_0) & \text{(center of the conic)} \\ \vec{u}(u_x, u_y) & \text{(eigenvector of } A_{33}\text{)} \end{cases}$$

So we can write a canonical equation:

$$a_{1,2} \stackrel{\text{def}}{=} \frac{x - x_0}{u_x} = \frac{y - y_0}{u_y}$$

Because a 2x2 matrix has 2 eigenvectors, we obtain 2 axes.

Vertices

For a general conic we can determine its vertices by calculating the intersection of the conic and its axes — in other words, by solving the system:

$$V \stackrel{\text{def}}{=} \begin{cases} e & \text{(axis)} \\ Q & \text{(the general equation of the conic)} \end{cases}$$

Tangents

Through a given point, P, there are generally two lines tangent to a conic. Expressing P as a column vector, p, the two points of tangency are the intersections of the conic with the line whose equation is

$$\mathbf{p}^T A_Q \mathbf{x} = 0$$

when P is on the conic, the line is the tangent there. When P is inside an ellipse, the line is the set of all points whose own associated line passes through P. This line is called the *polar* of the *pole* P with respect to the conic. Just as P uniquely determines its polar line (with respect to a given conic), so each line determines a unique P. This is thus an expression of geometric duality between points and lines in the plane.

As special cases, the centre of a conic is the pole of the line at infinity, and each asymptote of a hyperbola is a polar (a tangent) to one of its points at infinity. Using the theory of poles and polars, the problem of finding the four mutual tangents of two conics reduces to finding the intersection of two conics.

Reduced Equation

The *reduced equation of a conic section* is the equation of a conic section translated and rotated so that its centre lies in the centre of the coordinate system and its axes are parallel to the coordinate axes. This is equivalent to saying that the coordinates are moved to satisfy these properties.

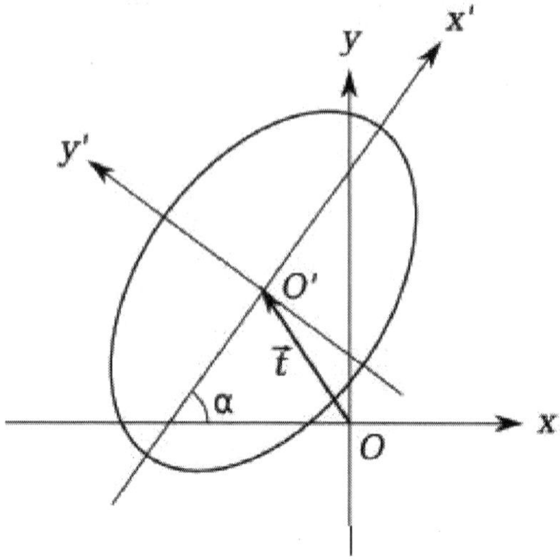

Figure: *If λ_1 and λ_2 are the eigenvalues of the matrix A_{33}, the reduced equation can be written as*

$$\lambda_1 x'^2 + \lambda_2 y'^2 + \frac{\det A_Q}{\det A_{33}} = 0$$

Dividing by $-\frac{\det A_Q}{\det A_{33}}$ we obtain a reduced canonical equation. For example, for an ellipse:

$$\frac{x'^2}{a^2} + \frac{y'^2}{b^2} = 1.$$

From here we get a and b.

The transformation of coordinates is given by:

$$T : RS(O,X,Y) \mapsto (O' = S, X', Y') \stackrel{\text{def}}{=} \begin{cases} \vec{t} & = \overrightarrow{OO'} = S \\ \alpha & = \arccos \dfrac{\vec{a}_1 \cdot \begin{pmatrix} 1 \\ 0 \end{pmatrix}}{|\vec{a}_1|} \end{cases}$$

As slice of quadratic form

The equation

$$Ax^2 + Bxy + Cy^2 + Dx + Ey + F = 0$$

can be rearranged by taking the affine linear part to the other side, yielding

$$Ax^2 + Bxy + Cy^2 = -(Dx + Ey + F).$$

In this form, a conic section is realised exactly as the intersection of the graph of the quadratic form $z = Ax^2 + Bxy + Cy^2$ and the plane $z = -(Dx + Ey + F)$. Parabolas and hyperbolas can be realised by a horizontal plane (D = E = 0), while ellipses require that the plane be slanted. Degenerate conics correspond to degenerate intersections, such as taking slices such as $z = -1$ of a positive-definite form.

Eccentricity in Terms of Parametres of the Quadratic Form

When the conic section is written algebraically as

$$Ax^2 + Bxy + Cy^2 + Dx + Ey + F = 0,$$

the eccentricity can be written as a function of the parametres of the quadratic equation. If $4AC = B^2$ the conic is a parabola and its eccentricity equals 1 (if it is non-degenerate). Otherwise, assuming the equation represents either a non-degenerate hyperbola or a non-degenerate, non-imaginary ellipse, the eccentricity is given by

$$e = \sqrt{\frac{2\sqrt{(A-C)^2 + B^2}}{\eta(A+C) + \sqrt{(A-C)^2 + B^2}}},$$

where $\eta = 1$ if the determinant of the 3×3 matrix is negative and $\eta = -1$ if that determinant is positive.

Applications

Conic sections are important in astronomy: the orbits of two massive objects that interact according to Newton's law of universal gravitation are conic sections if their common centre of mass is considered to be at rest. If they are bound together, they will both trace out ellipses; if they are moving apart, they will both follow parabolas or hyperbolas. In projective geometry, the conic sections in the projective plane are equivalent to each other up to projective transformations. For certain fossils in paleontology, understanding conic sections can help understand the three-dimensional shape of certain organisms.

Intersecting Two Conics

The solutions to a two second degree equations system in two variables may be seen as the coordinates of the intersections of two generic conic sections. In particular two conics may possess none, two or four possibly coincident intersection points. The best method of locating these solutions exploits the homogeneous matrix representation of conic sections, i.e. a 3×3 symmetric matrix which depends on six parametres.

The procedure to locate the intersection points follows these steps:

- Given the two conics C_1 and C_2 consider the pencil of conics given by their linear combination $\lambda C_1 + \mu C_2$
- Identify the homogeneous parametres (λ, μ) which corresponds to the degenerate conic of the pencil. This can be done by imposing that $\det(\lambda C_1 + \mu C_2) = 0$, which turns out to be the solution to a third degree equation.
- Given the degenerate conic C_0, identify the two, possibly coincident, lines constituting it
- Intersects each identified line with one of the two original conic; this step can be done efficiently using the dual conic representation of C_0
- The points of intersection will represent the solution to the initial equation system

Plane (Geometry)

In mathematics, a plane is a flat, two-dimensional surface. A plane is the two dimensional analogue of a point (zero-dimensions), a line (one-dimension) and a space (three-dimensions). Planes can arise as subspaces of some higher dimensional space, as with the walls of a room, or they may enjoy an independent existence in their own right, as in the setting of Euclidean geometry.

When working in two-dimensional Euclidean space, the definite article is used, the plane, to refer to the whole space. Many fundamental tasks in mathematics, geometry, trigonometry, graph theory and graphing are performed in two-dimensional space, or in other words, in the plane.

Euclidean Geometry

Euclidean geometry is a mathematical system attributed to the Alexandrian Greek mathematician Euclid, which he described in his textbook on geometry: the *Elements*. Euclid's method consists in assuming a small set of intuitively appealing axioms, and deducing many other propositions (theorems) from these. Although many of Euclid's results had been stated by earlier mathematicians, Euclid was the first to show how these propositions could fit into a comprehensive deductive and logical system. The *Elements* begins with plane geometry, still taught in secondary school as the first axiomatic system and the first examples of formal proof. It goes on to the solid geometry of three dimensions. Much of the *Elements* states results of what are now called algebra and number theory, couched in geometrical language.

For over two thousand years, the adjective "Euclidean" was unnecessary because no other sort of geometry had been conceived. Euclid's axioms seemed so intuitively obvious that any theorem proved from them was deemed true in an absolute, often metaphysical, sense. Today, however, many other self-consistent non-Euclidean geometries are known, the first ones having been discovered in the early 19th century. An implication of Einstein's theory of general relativity is that Euclidean space is a good approximation to the properties of physical space only where the gravitational field is weak.

The Elements

The *Elements* are mainly a systematisation of earlier knowledge of geometry. Its superiority over earlier treatments was rapidly recognised, with the result that there was little interest in preserving the earlier ones, and they are now nearly all lost.

Books I–IV and VI discuss plane geometry. Many results about plane figures are proved, e.g., *If a triangle has two equal angles, then the sides subtended by the angles are equal.* The Pythagorean theorem is proved.

Books V and VII–X deal with number theory, with numbers treated geometrically via their representation as line segments with various lengths. Notions such as prime numbers and rational and irrational numbers are introduced. The infinitude of prime numbers is proved.

Books XI–XIII concern solid geometry. A typical result is the 1:3 ratio between the volume of a cone and a cylinder with the same height and base.

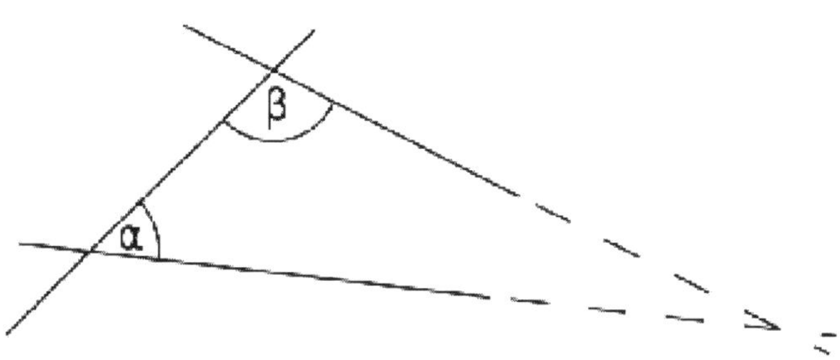

The Parallel Postulate: If two lines intersect a third in such a way that the sum of the inner angles on one side is less than two right angles, then the two lines inevitably must intersect each other on that side if extended far enough.

Axioms

Euclidean geometry is an axiomatic system, in which all theorems ("true statements") are derived from a small number of axioms. Near the beginning of the first book of the *Elements*, Euclid gives five postulates (axioms) for plane geometry, stated in terms of constructions (as translated by Thomas Heath):

"Let the following be postulated":

1. "To draw a straight line from any point to any point."
2. "To produce [extend] a finite straight line continuously in a straight line."
3. "To describe a circle with any centre and distance [radius]."
4. "That all right angles are equal to one another."
5. *The parallel postulate*: "That, if a straight line falling on two straight lines make the interior angles on the same side less than two right angles, the two straight lines, if produced indefinitely, meet on that side on which are the angles less than the two right angles."

Although Euclid's statement of the postulates only explicitly asserts the existence of the constructions, they are also taken to be unique.

The *Elements* also include the following five "common notions":

1. Things that are equal to the same thing are also equal to one another.
2. If equals are added to equals, then the wholes are equal.
3. If equals are subtracted from equals, then the remainders are equal.
4. Things that coincide with one another equal one another.
5. The whole is greater than the part.

Parallel Postulate

To the ancients, the parallel postulate seemed less obvious than the others. Euclid himself seems to have considered it as being qualitatively different from the others, as evidenced by the organisation of the *Elements*: the first 28 propositions he presents are those that can be proved without it. Many alternative axioms can be formulated that have the same logical consequences as the parallel postulate. For example Playfair's axiom states:

In a plane, through a point not on a given straight line, at most one line can be drawn that never meets the given line.

Methods of Proof

Euclidean geometry is *constructive*. Postulates 1, 2, 3, and 5 assert the existence and uniqueness of certain geometric figures, and these assertions are of a constructive nature: that is, we are not only told that certain things exist, but are also given methods for creating them with no more than a compass and an unmarked straightedge. In this sense, Euclidean geometry is more concrete than many modern axiomatic systems such as set theory, which often assert the existence of objects without saying how to construct them, or even assert the existence of objects that cannot be constructed within the theory. Strictly speaking, the lines on paper are *models* of the objects defined within the formal system, rather than instances of those objects. For example a Euclidean straight line has no width, but any real drawn line will. Though nearly all modern mathematicians consider nonconstructive methods just as sound as constructive ones, Euclid's constructive proofs often supplanted fallacious nonconstructive ones—e.g., some of the Pythagoreans' proofs that involved irrational numbers, which usually required a statement such as "Find the greatest common measure of ..."

Euclid often used proof by contradiction. Euclidean geometry also allows the method of superposition, in which a figure is transferred to another point in space. For example, proposition I.4, side-angle-side congruence of triangles, is proved by moving one of the two triangles so that one of its sides coincides with the other triangle's equal side, and then proving that the other sides coincide as well. Some modern treatments add a sixth postulate, the rigidity of the triangle, which can be used as an alternative to superposition.

System of Measurement and Arithmetic

Euclidean geometry has two fundamental types of measurements: angle and distance. The angle scale is absolute, and Euclid uses the right angle as his basic unit, so that, e.g., a 45-degree angle would be referred to as half of a right angle. The distance scale is relative; one arbitrarily picks a line segment with a certain length as the unit, and other distances are expressed in relation to it.

A line in Euclidean geometry is a model of the real number line. A line segment is a part of a line that is bounded by two end points, and contains every point on the line between its end points. Addition is represented by a construction in which one line segment is copied onto the end of another line segment to extend its length, and similarly for subtraction.

Measurements of area and volume are derived from distances. For example, a rectangle with a width of 3 and a length of 4 has an area that represents the product, 12. Because this geometrical interpretation of multiplication was limited to three dimensions, there was no direct way of interpreting the product of four or more numbers, and Euclid avoided such products, although they are implied, e.g., in the proof of book IX, proposition 20.

Euclid refers to a pair of lines, or a pair of planar or solid figures, as "equal" (4σïὸ) if their lengths, areas, or volumes are equal, and similarly for angles. The stronger term "congruent" refers to the idea that an entire figure is the same size and shape as another figure. Alternatively, two figures are congruent if one can be moved on top of the other so that it matches up with it exactly. (Flipping it over is allowed.) Thus, for example, a 2 × 6 rectangle and a 3x4 rectangle are equal but not congruent, and the letter R is congruent to its mirror image. Figures that would be congruent except for their differing sizes are referred to as similar.

Notation and Terminology

Naming of Points and Figures

Points are customarily named using capital letters of the alphabet. Other figures, such as lines, triangles, or circles, are named by listing a sufficient number of points to pick them out unambiguously from the relevant figure, e.g., triangle ABC would typically be a triangle with vertices at points A, B, and C.

Complementary and Supplementary Angles

Angles whose sum is a right angle are called complementary. Complementary angles are formed when one or more rays share the same vertex and are pointed in a direction that is in between the two original rays that form the right angle. The number of rays in between the two original rays are infinite. Those whose sum is a straight angle are supplementary. Supplementary angles are formed when one or more rays share the same vertex and are pointed in a direction that in between the two original rays that form the straight angle (180 degrees). The number of rays in between the two original rays are infinite like those possible in the complementary angle.

Modern Versions of Euclid's Notation

In modern terminology, angles would normally be measured in degrees or radians. Modern school textbooks often define separate figures called lines (infinite), rays (semi-infinite), and line segments (of finite length). Euclid, rather than discussing a ray as an object that extends to infinity in one direction, would normally use locutions such as "if the line is extended to a sufficient length," although he occasionally referred to "infinite lines." A "line" in Euclid could be either straight or curved, and he used the more specific term "straight line" when necessary.

Bridge of Asses

The Bridge of Asses (*Pons Asinorum*) states that *in isosceles triangles the angles at the base equal one another, and, if the equal straight lines are produced further, then the angles under the base equal one another*. Its name may be attributed to its frequent role as the first real test in the *Elements* of the intelligence of the reader and as a bridge to the harder propositions that followed. It might also be so named because of the geometrical figure's resemblance to a steep bridge that only a sure-footed donkey could cross.

Congruence of Triangles

Triangles are congruent if they have all three sides equal (SSS), two sides and the angle between them equal (SAS), or two angles and a side equal (ASA) (Book I, propositions 4, 8, and 26). (Triangles with three equal angles are generally similar, but not necessarily congruent. Also, triangles with two equal sides and an adjacent angle are not necessarily equal.)

Sum of the Angles of a Triangle Acute, Obtuse and Right Angle Limits

The sum of the angles of a triangle is equal to a straight angle (180 degrees). This causes an equilateral triangle to have 3 interior angles of 60 degrees. Also, it causes every triangle to have at least 2 acute angles and up to 1 obtuse or right angle.

Pythagorean Theorem

The celebrated Pythagorean theorem (book I, proposition 47) states that in any right triangle, the area of the square whose side is the hypotenuse (the side opposite the right angle) is equal to the sum of the areas of the squares whose sides are the two legs (the two sides that meet at a right angle).

Thales' Theorem

Thales' theorem, named after Thales of Miletus states that if A, B, and C are points on a circle where the line AC is a diametre of the circle, then the angle ABC is a right angle. Cantor supposed that Thales proved his theorem by means of Euclid book I, prop 32 after the manner of Euclid book III, prop 31. Tradition has it that Thales sacrificed an ox to celebrate this theorem.

As a Description of the Structure of Space

Euclid believed that his axioms were self-evident statements about physical reality. Euclid's proofs depend upon assumptions perhaps not obvious in Euclid's fundamental axioms, in particular that certain movements of figures do not change their geometrical properties such as the lengths of sides and interior angles, the so-called *Euclidean motions*, which include translations and rotations of figures. Taken as a physical description of space, postulate 2 (extending a line) asserts that space does not have holes or boundaries (in other words, space is homogeneous and unbounded); postulate 4 (equality of right angles) says that space is isotropic and figures may be moved to any location while maintaining congruence; and postulate 5 (the parallel postulate) that space is flat (has no intrinsic curvature). As discussed in more detail below, Einstein's theory of relativity significantly modifies this view. The ambiguous character of the axioms as originally formulated by Euclid makes it possible for different commentators to disagree about some of their other implications for the structure of space, such as whether or not it is infinite and what its topology is. Modern, more rigorous reformulations of the system typically aim for a cleaner separation of these issues. Interpreting Euclid's axioms in the spirit of this more modern approach, axioms 1-4 are consistent with either infinite or finite space (as in elliptic geometry), and all five axioms are consistent with a variety of topologies (e.g., a plane, a cylinder, or a torus for two-dimensional Euclidean geometry).

Later Work

Archimedes and Apollonius

Archimedes (ca. 287 BCE – ca. 212 BCE), a colourful figure about whom many historical anecdotes are recorded, is remembered along with Euclid as one of the greatest of ancient mathematicians. Although the foundations of his work were put in place by Euclid, his work, unlike Euclid's, is believed to have been entirely original. He proved equations for the volumes and areas of various figures in two and three dimensions, and enunciated the Archimedean property of finite numbers.

Apollonius of Perga (ca. 262 BCE–ca. 190 BCE) is mainly known for his investigation of conic sections.

17th Century: Descartes

René Descartes (1596–1650) developed analytic geometry, an alternative method for formalising geometry. In this approach, a point is represented by its Cartesian (x, y) coordinates, a line is represented by its equation, and so on. In Euclid's original approach, the Pythagorean theorem follows from Euclid's axioms. In the Cartesian approach, the axioms are the axioms of algebra, and the equation expressing the Pythagorean theorem is then a definition of one of the terms in Euclid's axioms, which are now considered theorems. The equation

$$|PQ|=\sqrt{(p-r)^2+(q-s)^2}$$

defining the distance between two points $P = (p, q)$ and $Q = (r, s)$ is then known as the *Euclidean metric*, and other metrics define non-Euclidean geometries.

In terms of analytic geometry, the restriction of classical geometry to compass and straightedge constructions means a restriction to first- and second-order equations, e.g., $y = 2x + 1$ (a line), or $x^2 + y^2 = 7$ (a circle). Also in the 17th century, Girard Desargues, motivated by the theory of perspective, introduced the concept of idealised points, lines, and planes at infinity. The result can be considered as a type of generalised geometry, projective geometry, but it can also be used to produce proofs in ordinary Euclidean geometry in which the number of special cases is reduced.

18th Century

Geometres of the 18th century struggled to define the boundaries of the Euclidean system. Many tried in vain to prove the fifth postulate from the first four. By 1763 at least 28 different proofs had been published, but all were found incorrect.

Leading up to this period, geometres also tried to determine what constructions could be accomplished in Euclidean geometry. For example, the problem of trisecting an angle with a compass and straightedge is one that naturally occurs within the theory, since the axioms refer to constructive operations that can be carried out with those tools. However, centuries of efforts failed to find a solution to this problem, until Pierre Wantzel published a proof in 1837 that such a construction was impossible. Other constructions that were proved impossible include doubling the cube and squaring the circle. In the case of doubling the cube, the impossibility of the construction originates from the fact that the compass and straightedge method involve first- and second-order equations, while doubling a cube requires the solution of a third-order equation.

Euler discussed a generalisation of Euclidean geometry called affine geometry, which retains the fifth postulate unmodified while weakening postulates three and four in a way that eliminates the notions of angle (whence right triangles become meaningless) and of equality of length of line segments in general (whence circles become meaningless) while retaining the notions of parallelism as an equivalence relation between lines, and equality of length of parallel line segments (so line segments continue to have a midpoint).

19th Century and Non-euclidean Geometry

In the early 19th century, Carnot and Möbius systematically developed the use of signed angles and line segments as a way of simplifying and unifying results.

The century's most significant development in geometry occurred when, around 1830, János Bolyai and Nikolai Ivanovich Lobachevsky separately published work on non-Euclidean geometry, in which the parallel postulate is not valid. Since non-Euclidean geometry is provably relatively consistent with Euclidean geometry, the parallel postulate cannot be proved from the other postulates.

In the 19th century, it was also realised that Euclid's ten axioms and common notions do not suffice to prove all of theorems stated in the *Elements*. For example, Euclid assumed implicitly that any line contains at least two points, but this assumption cannot be proved from the other axioms, and therefore must be an axiom itself. The very first geometric proof in the *Elements,* shown in the figure above, is that any line segment is part of a triangle; Euclid constructs this in the usual way, by drawing circles around both endpoints and taking their intersection as the third vertex. His axioms, however, do not guarantee that the circles actually intersect, because they do not assert the geometrical property of continuity, which in Cartesian terms is equivalent to the completeness property of the real numbers.

Starting with Moritz Pasch in 1882, many improved axiomatic systems for geometry have been proposed, the best known being those of Hilbert, George Birkhoff, and Tarski.

20th Century and General Relativity

Einstein's theory of general relativity shows that the true geometry of spacetime is not Euclidean geometry. For example, if a triangle is constructed out of three rays of light, then in general the interior angles do not add up to 180 degrees due to gravity. A relatively weak gravitational field, such as the Earth's or the sun's, is represented by a metric that is approximately, but not exactly, Euclidean. Until the 20th century, there was no technology capable of detecting the deviations from Euclidean geometry, but Einstein predicted that such deviations would exist. They were later verified by observations such as the slight bending of starlight by the Sun during a solar eclipse in 1919, and such considerations are now an integral part of the software that runs the GPS system. It is possible to object to this interpretation of general relativity on the grounds that light rays might be improper physical models of Euclid's lines, or that relativity could be rephrased so as to avoid the geometrical interpretations. However, one of the consequences of Einstein's theory is that there is no possible physical test that can distinguish between a beam of light as a model of a geometrical line and any other physical model. Thus, the only logical possibilities are to accept non-Euclidean geometry as physically real, or to reject the entire notion of physical tests of the axioms of geometry, which can then be imagined as a formal system without any intrinsic real-world meaning.

Chapter

6

Treatment of Infinity

Infinite Objects

Euclid sometimes distinguished explicitly between "finite lines" (e.g., Postulate 2) and "infinite lines" (book I, proposition 12). However, he typically did not make such distinctions unless they were necessary. The postulates do not explicitly refer to infinite lines, although for example some commentators interpret postulate 3, existence of a circle with any radius, as implying that space is infinite.

The notion of infinitesimally small quantities had previously been discussed extensively by the Eleatic School, but nobody had been able to put them on a firm logical basis, with paradoxes such as Zeno's paradox occurring that had not been resolved to universal satisfaction. Euclid used the method of exhaustion rather than infinitesimals.

Later ancient commentators such as Proclus (410–485 CE) treated many questions about infinity as issues demanding proof and, e.g., Proclus claimed to prove the infinite divisibility of a line, based on a proof by contradiction in which he considered the cases of even and odd numbers of points constituting it.

At the turn of the 20th century, Otto Stolz, Paul du Bois-Reymond, Giuseppe Veronese, and others produced controversial work on non-Archimedean models of Euclidean geometry, in which the distance between two points may be infinite or infinitesimal, in the Newton–Leibniz sense. Fifty years later, Abraham Robinson provided a rigorous logical foundation for Veronese's work.

Infinite Processes

One reason that the ancients treated the parallel postulate as less certain than the others is that verifying it physically would require us to inspect two lines to check that they never intersected, even at some very distant point, and this inspection could potentially take an infinite amount of time.

The modern formulation of proof by induction was not developed until the 17th century, but some later commentators consider it implicit in some of Euclid's proofs, e.g., the proof of the infinitude of primes.

Supposed paradoxes involving infinite series, such as Zeno's paradox, predated Euclid. Euclid avoided such discussions, giving, for example, the expression for the partial sums of the geometric series in IX.35 without commenting on the possibility of letting the number of terms become infinite.

Logical Basis

Classical Logic

Euclid frequently used the method of proof by contradiction, and therefore the traditional presentation of Euclidean geometry assumes classical logic, in which every proposition is either true or false, i.e., for any proposition *P*, the proposition "*P* or not *P*" is automatically true.

Modern Standards of Rigor

Placing Euclidean geometry on a solid axiomatic basis was a preoccupation of mathematicians for centuries. The role of primitive notions, or undefined concepts, was clearly put forward by Alessandro Padoa of the Peano delegation at the 1900 Paris conference:

> *...when we begin to formulate the theory, we can imagine that the undefined symbols are* completely devoid of meaning *and that the unproved propositions are simply* conditions *imposed upon the undefined symbols.*

Then, the *system of ideas* that we have initially chosen is simply *one interpretation* of the undefined symbols; but..this interpretation can be ignored by the reader, who is free to replace it in his mind by *another interpretation..* that satisfies the conditions...

Logical questions thus become completely independent of *empirical* or *psychological* questions...

The system of undefined symbols can then be regarded as the *abstraction* obtained from the *specialised theories* that result when...the system of undefined symbols is successively replaced by each of the interpretations...

—Padoa, *Essai d'une théorie algébrique des nombre entiers, avec une Introduction logique à une théorie déductive qulelconque*

That is, mathematics is context-independent knowledge within a hierarchical framework. As said by Bertrand Russell:

> *If our hypothesis is about* anything, *and not about some one or more particular things, then our deductions constitute mathematics. Thus, mathematics may be defined as the subject in which we never know what we are talking about, nor whether what we are saying is true.*

—Bertrand Russell, *Mathematics and the metaphysicians*

Such foundational approaches range between foundationalism and formalism.

Axiomatic Formulations

Geometry is the science of correct reasoning on incorrect figures.

—George Polyá, *How to Solve It, p. 208*

- *Euclid's axioms:* In his dissertation to Trinity College, Cambridge, Bertrand Russell summarised the changing role of Euclid's geometry in the minds of philosophers up to that time. It was a conflict between certain knowledge, independent of experiment, and empiricism, requiring experimental input. This issue became clear as it was discovered that the parallel postulate was not necessarily valid and its applicability was an empirical matter, deciding whether the applicable geometry was Euclidean or non-Euclidean.
- *Hilbert's axioms:* Hilbert's axioms had the goal of identifying a *simple* and *complete* set of *independent* axioms from which the most important geometric theorems could be deduced. The outstanding objectives were to make Euclidean geometry rigorous (avoiding hidden assumptions) and to make clear the ramifications of the parallel postulate.
- *Birkhoff's axioms:* Birkhoff proposed four postulates for Euclidean geometry that can be confirmed experimentally with scale and protractor. The notions of *angle* and *distance* become primitive concepts.

- *Tarski's axioms:* Tarski (1902–1983) and his students defined *elementary* Euclidean geometry as the geometry that can be expressed in first-order logic and does not depend on set theory for its logical basis, in contrast to Hilbert's axioms, which involve point sets. Tarski proved that his axiomatic formulation of elementary Euclidean geometry is consistent and complete in a certain sense: there is an algorithm that, for every proposition, can be shown either true or false. (This doesn't violate Gödel's theorem, because Euclidean geometry cannot describe a sufficient amount of arithmetic for the theorem to apply.) This is equivalent to the decidability of real closed fields, of which elementary Euclidean geometry is a model.

Constructive Approaches and Pedagogy

The process of abstract axiomatisation as exemplified by Hilbert's axioms reduces geometry to theorem proving or predicate logic. In contrast, the Greeks used construction postulates, and emphasised problem solving. For the Greeks, constructions are more primitive than existence propositions, and can be used to prove existence propositions, but not *vice versa*. To describe problem solving adequately requires a richer system of logical concepts. The contrast in approach may be summarised:

- *Axiomatic proof:* Proofs are deductive derivations of propositions from primitive premises that are 'true' in some sense. The aim is to justify the proposition.
- *Analytic proof:* Proofs are non-deductive derivations of hypothesis from problems. The aim is to find hypotheses capable of giving a solution to the problem. One can argue that Euclid's axioms were arrived upon in this manner. In particular, it is thought that Euclid felt the parallel postulate was forced upon him, as indicated by his reluctance to make use of it, and his arrival upon it by the method of contradiction.

Andrei Nicholaevich Kolmogorov proposed a problem solving basis for geometry. This work was a precursor of a modern formulation in terms of constructive type theory. This development has implications for pedagogy as well.

If proof simply follows conviction of truth rather than contributing to its construction and is only experienced as a demonstration of something already known to be true, it is likely to remain meaningless and purposeless in the eyes of students.

—Celia Hoyles, *The curricular shaping of students' approach to proof*

Planes Embedded in 3-dimensional Euclidean Space

This section is specifically concerned with planes embedded in three dimensions: specifically, in R^3.

Properties

In three-dimensional Euclidean space, we may exploit the following facts that do not hold in higher dimensions:

- Two planes are either parallel or they intersect in a line.
- A line is either parallel to a plane, intersects it at a single point, or is contained in the plane.
- Two lines perpendicular to the same plane must be parallel to each other.
- Two planes perpendicular to the same line must be parallel to each other.

Planes in Various Areas of Mathematics

In addition to its familiar geometric structure, with isomorphisms that are isometries with respect to the usual inner product, the plane may be viewed at various other levels of abstraction. Each level of abstraction corresponds to a specific category.

At one extreme, all geometrical and metric concepts may be dropped to leave the topological plane, which may be thought of as an idealised homotopically trivial infinite rubber sheet, which retains a notion of proximity, but has no distances. The topological plane has a concept of a linear path, but no concept of a straight line. The topological plane, or its equivalent the open disc, is the basic topological neighbourhood used to construct surfaces (or 2-manifolds) classified in low-dimensional topology. Isomorphisms of the topological plane are all continuous bijections. The topological plane is the natural context for the branch of graph theory that deals with planar graphs, and results such as the four colour theorem.

The plane may also be viewed as an affine space, whose isomorphisms are combinations of translations and non-singular linear maps. From this viewpoint there are no distances, but collinearity and ratios of distances on any line are preserved.

Differential geometry views a plane as a 2-dimensional real manifold, a topological plane which is provided with a differential structure. Again in this case, there is no notion of distance, but there is now a concept of smoothness of maps, for example a differentiable or smooth path (depending on the type of differential structure applied). The isomorphisms in this case are bijections with the chosen degree of differentiability.

In the opposite direction of abstraction, we may apply a compatible field structure to the geometric plane, giving rise to the complex plane and the major area of complex analysis. The complex field has only two isomorphisms that leave the real line fixed, the identity and conjugation. In the same way as in the real case, the plane may also be viewed as the simplest, one-dimensional (over the complex numbers) complex manifold, sometimes called the complex line. However, this viewpoint contrasts sharply with the case of the plane as a 2-dimensional real manifold. The isomorphisms are all conformal bijections of the complex plane, but the only possibilities are maps that correspond to the composition of a multiplication by a complex number and a translation.

In addition, the Euclidean geometry (which has zero curvature everywhere) is not the only geometry that the plane may have. The plane may be given a spherical geometry by using the stereographic projection. This can be thought of as placing a sphere on the plane (just like a ball on the floor), removing the top point, and projecting the sphere onto the plane from this point). This is one of the projections that may be used in making a flat map of part of the Earth's surface. The resulting geometry has constant positive curvature.

Alternatively, the plane can also be given a metric which gives it constant negative curvature giving the hyperbolic plane. The latter possibility finds an application in the theory of special relativity in the simplified case where there are two spatial dimensions and one time dimension. (The hyperbolic plane is a timelike hypersurface in three-dimensional Minkowski space.)

Topological and Differential Geometric Notions

The one-point compactification of the plane is homeomorphic to a sphere; the open disk is homeomorphic to a sphere with the "north pole" missing; adding that point completes the (compact) sphere. The result of this compactification is a manifold referred to as the Riemann sphere or the complex projective line. The projection from the Euclidean plane to a sphere without a point is a diffeomorphism and even a conformal map. The plane itself is homeomorphic (and diffeomorphic) to an open disk. For the Lobachevsky plane such diffeomorphism is conformal, but for the Euclidean plane it is not.

Line (Geometry)

The notion of line or straight line was introduced by ancient mathematicians to represent straight objects with negligible width and depth. Lines are an idealisation of such objects. Thus, until seventeenth century, lines were defined like this: "The line is the first species of quantity, which has only one dimension, namely length, without any width nor depth, and is nothing else than the flow or run of the point which [...] will leave from its imaginary moving some vestige in length, exempt of any width. [...] The straight line is that which is equally extended between its points"

Euclid described a line as "breadthless length", and introduced several postulates as basic unprovable properties from which he constructed the geometry, which is now called Euclidean geometry to avoid confusion with other geometries which have been introduced since the end of nineteenth century (such as non-Euclidean geometry, projective geometry, and affine geometry).

In modern mathematics, given the multitude of geometries, the concept of a line is closely tied to the way the geometry is described. For instance, in analytic geometry, a line in the plane is often defined as the set of points whose coordinates satisfy a given linear equation, but in a more abstract setting, such as incidence geometry, a line may be an independent object, distinct from the set of points which lie on it.

When a geometry is described by a set of axioms, the notion of a line is usually left undefined (a so-called primitive object). The properties of lines are then determined by the axioms which refer to them. One advantage to this approach is the flexibility it gives to users of the geometry.

Thus in differential geometry a line may be interpreted as a geodesic (shortest path between points), while in some projective geometries a line is a 2-dimensional vector space (all linear combinations of two independent vectors). This flexibility also extends beyond mathematics and, for example, permits physicists to think of the path of a light ray as being a line.

A line segment is a part of a line that is bounded by two distinct end points and contains every point on the line between its end points. Depending on how the line segment is defined, either of the two end points may or may not be part of the line segment. Two or more line segments may have some of the same relationships as lines, such as being parallel, intersecting, or skew.

Ray

If the concept of "order" of points of a line is defined, a *ray*, or half-line, may be defined as well. A ray is part of a line which is finite in one direction, but infinite in the other. It can be defined by two points, the initial point, A, and one other, B. The ray is all the points in the line segment between A and B together with all points, C, on the line through A and B such that the points appear on the line in the order A, B, C.

In topology, a ray in a space X is a continuous embedding $R^+ \to X$. It is used to define the important concept of end of the space.

Vector Equation

The vector equation of the line through points A and B is given by r = OA + λAB (where λ is a scalar multiple).

If a is vector OA and b is vector OB, then the equation of the line can be written: r = a + λ(b - a).

A ray starting at point A is described by limiting $\lambda \geq 0$.

Collinear Points

Three points are said to be collinear if they lie on the same line. In Euclidean space, this is the degenerate condition where three points do *not* determine a plane.

In coordinate geometry, the points $X = (x_1, x_2, ...)$, $Y = (y_1, y_2, ...)$, and $Z = (z_1, z_2, ...)$ are collinear if the matrix

$$\begin{bmatrix} 1 & x_1 & x_2 & \cdots \\ 1 & y_1 & y_2 & \cdots \\ 1 & z_1 & z_2 & \cdots \end{bmatrix}$$

has a rank less than 3. In particular, for three points in the plane, above matrix is square and the points are collinear if its determinant is zero.

In the geometries where the line is not a primitive notion, as may be the case in some synthetic geometries, another definition of collinearity is needed.

When the distance $d(a, b)$ between two points a and b is a primitive notion, the collinearity between three points may be defined by:

The points a, b and c are collinear if and only if $d(x,a) = d(c, a)$ and $d(x, b) = d(c, b)$ implies $x = c$.

In Euclidean geometry this property is true, because the point which is symmetric to *c* with respect to the line defined by *a* and *b* satisfies the equalities on the left of "implies", and this point is equal to *c* if and only if *c* is on the line.

Euclidean Space

In Euclidean space, R^n (and analogously in every other affine space), the line *L* passing through two different points *a* and *b* is the subset

$$L = \{(1-t)a+tb\ t \in \square\}$$

The direction of the line is that of the vector *b-a*. Different choices of *a* and *b* can yield the same line.

Projective Geometry

In projective geometry, a line is similar to that in Euclidean geometry but has slightly different properties. In many models of projective geometry, the idea of the line rarely conforms to the notion of the "straight curve" as it is visualised in Euclidean geometry. Elliptic geometry is a typical example of when this happens.

Geodesics

The "straightness" of a line, interpreted as the property that it minimises distances between its points, can be generalised and leads to the concept of geodesics in metric spaces.

Rotation (Mathematics)

In geometry and linear algebra, a rotation is a transformation in a plane or in space that describes the motion of a rigid body around a fixed point. A rotation is different from a translation, which has no fixed points, and from a reflection, which "flips" the bodies it is transforming. A rotation and the above-mentioned transformations are isometries; they leave the distance between any two points unchanged after the transformation.

It is important to know the frame of reference when considering rotations, as all rotations are described relative to a particular frame of reference. In general for any orthogonal transformation on a body in a coordinate system there is an inverse transformation which if applied to the frame of reference results in the body being at the same coordinates. For example in two dimensions rotating a body clockwise about a point keeping the axes fixed is equivalent to rotating the axes counterclockwise about the same point while the body is kept fixed.

Circle Group

In mathematics, the circle group, denoted by T, is the multiplicative group of all complex numbers with absolute value 1, i.e., the unit circle in the complex plane.

$$\mathbb{T} = \{z \in \mathbb{C} : |z| = 1\}.$$

The circle group forms a subgroup of $C^{\times}$, the multiplicative group of all nonzero complex numbers. Since $C^{\times}$ is abelian, it follows that T is as well. The circle group is also the group U(1) of 1×1 unitary matrices; these act on the complex plane by rotation about the origin. The circle group can be parametrised by the angle θ of rotation by

$$\theta \mapsto z = e^{i\theta} = \cos\theta + i\sin\theta.$$

This is the exponential map for the circle group. The circle group plays a central role in Pontryagin duality, and in the theory of Lie groups. The notation T for the circle group stems from the fact that T^n (the direct product of T with itself n times) is geometrically an n-torus. The circle group is then a 1-torus.

Elementary Introduction

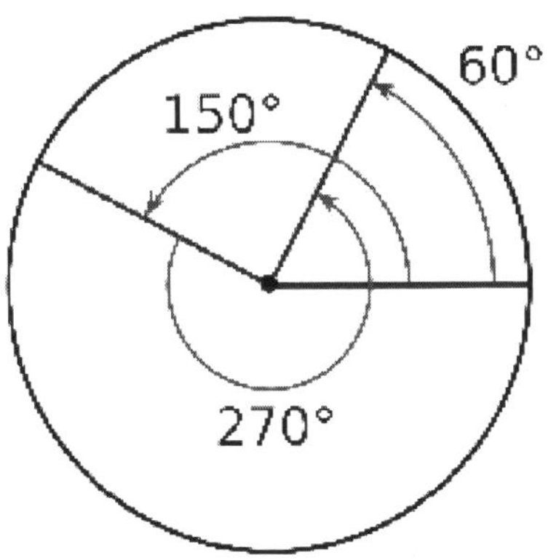

Figure: *Multiplication on the circle group is equivalent to addition of angles*

One way to think about the circle group is that it describes how to add *angles*, where only angles between 0° and 360° are permitted. For example, the diagram illustrates how to add 150° to 270°. The answer should be 150° + 270° = 420°, but when thinking in terms of the circle group, we need to "forget" the fact that we have wrapped once around the circle. Therefore we adjust our answer by 360° which gives 420° = 60° (mod 360°).

Another description is in terms of ordinary addition, where only numbers between 0 and 1 are allowed (with 1 corresponding to a full rotation). To achieve this, we might need to throw away digits occurring before the decimal point. For example, when we work out 0.784 + 0.925 + 0.446, the answer should be 2.155, but we throw away the leading 2, so the answer (in the circle group) is just 0.155.

Topological and Analytic Structure

The circle group is more than just an abstract algebraic object. It has a natural topology when regarded as a subspace of the complex plane. Since multiplication and inversion are continuous functions on $C^{\times}$, the circle group has the structure of a topological group. Moreover, since the unit circle is a closed subset of the complex plane, the circle group is a closed subgroup of $C^{\times}$ (itself regarded as a topological group).

One can say even more. The circle is a 1-dimensional real manifold and multiplication and inversion are real-analytic maps on the circle. This gives the circle group the structure of a one-parametre group, an instance of a Lie group. In fact, up to isomorphism, it is the unique 1-dimensional compact, connected Lie group. Moreover, every n-dimensional compact, connected, abelian Lie group is isomorphic to T^n.

Isomorphisms

The circle group shows up in a variety of forms in mathematics. We list some of the more common forms here. Specifically, we show that

$$\mathbb{T} \cong U(1) \cong □ \,/\, □ \cong SO(2).$$

Note that the slash (/) denotes here quotient group.

The set of all 1×1 unitary matrices clearly coincides with the circle group; the unitary condition is equivalent to the condition that its element have absolute value 1. Therefore, the circle group is canonically isomorphic to U(1), the first unitary group.

The exponential function gives rise to a group homomorphism exp : R → T from the additive real numbers R to the circle group T via the map

$$\theta \mapsto e^{i\theta} = \cos\theta + i\sin\theta.$$

The last equality is Euler's formula. The real number θ corresponds to the angle on the unit circle as measured from the positive x-axis. That this map is a homomorphism follows from the fact that the multiplication of unit complex numbers corresponds to addition of angles:

$$e^{i\theta_1}e^{i\theta_2} = e^{i(\theta_1+\theta_2)}.$$

This exponential map is clearly a surjective function from R to T. It is not, however, injective. The kernel of this map is the set of all integer multiples of 2π. By the first isomorphism theorem we then have that

$$\mathbb{T} \cong \mathbb{R} / 2\pi \mathbb{Z}$$

After rescaling we can also say that T is isomorphic to R/Z.

If complex numbers are realised as 2 × 2 real matrices, the unit complex numbers correspond to 2 × 2 orthogonal matrices with unit determinant. Specifically, we have

$$e^{i\theta} \leftrightarrow \begin{bmatrix} \cos\theta & -\sin\theta \\ \sin\theta & \cos\theta \end{bmatrix}.$$

The circle group is therefore isomorphic to the special orthogonal group SO(2). This has the geometric interpretation that multiplication by a unit complex number is a proper rotation in the complex plane, and every such rotation is of this form.

Properties

Every compact Lie group G of dimension > 0 has a subgroup isomorphic to the circle group. That means that, thinking in terms of symmetry, a compact symmetry group acting *continuously* can be expected to have one-parametre circle subgroups acting; the consequences in physical systems are seen for example at rotational invariance, and spontaneous symmetry breaking.

The circle group has many subgroups, but its only proper closed subgroups consist of roots of unity: For each integer $n > 0$, the nth roots of unity form a cyclic group of order n, which is unique up to isomorphism.

Representations

The representations of the circle group are easy to describe. It follows from Schur's lemma that the irreducible complex representations of an abelian group are all 1-dimensional. Since the circle group is compact, any representation ρ : T → GL(1, C) E– C$^{\times}$, must take values in U(1) E– T. Therefore, the irreducible representations of the circle group are just the homomorphisms from the circle group to itself. Every such homomorphism is of the form

$$\phi_n(e^{i\theta}) = e^{in\theta}, \qquad n \in \mathbb{Z}.$$

These representations are all inequivalent. The representation φ_{-n} is conjugate to φ_n,

$$\phi_{-n} = \overline{\phi_n}.$$

These representations are just the characters of the circle group. The character group of T is clearly an infinite cyclic group generated by φ_1:

$$\mathrm{Hom}(\mathbb{T},\mathbb{T}) \cong \mathbb{Z}.$$

The irreducible real representations of the circle group are the trivial representation (which is 1-dimensional) and the representations

$$\rho_n(e^{i\theta}) = \begin{bmatrix} \cos n\theta & -\sin n\theta \\ \sin n\theta & \cos n\theta \end{bmatrix}, \quad n \in \mathbb{Z}^+,$$

taking values in SO(2). Here we only have positive integers n since the representation ρ_{-n} is equivalent to ρ_n.

Rotation Group SO(3)

In mechanics and geometry, the 3D rotation group is the group of all rotations about the origin of three-dimensional Euclidean space R^3 under the operation of composition. By definition, a rotation about the origin is a linear transformation that preserves length of vectors (it is an isometry) and preserves orientation (i.e. *handedness*) of space. A length-preserving transformation which reverses orientation is an improper rotation, that is a reflection or more generally a rotoinversion.

Composing two rotations results in another rotation; every rotation has a unique inverse rotation; and the identity map satisfies the definition of a rotation. Owing to the above properties (along with the associative property, which rotations obey,) the set of all rotations is a group under composition. Moreover, the rotation group has a natural manifold structure for which the group operations are smooth; so it is in fact a Lie group. The rotation group is often denoted SO(3) for reasons explained below.

Length and Angle

Besides just preserving length, rotations also preserve the angles between vectors. This follows from the fact that the standard dot product between two vectors u and v can be written purely in terms of length:

Besides just preserving length, rotations also preserve the angles between vectors. This follows from the fact that the standard dot product between two vectors u and v can be written purely in terms of length:

$$\mathbf{u}\cdot\mathbf{v} = \tfrac{1}{2}\left(\|\mathbf{u}+\mathbf{v}\|^2 - \|\mathbf{u}\|^2 - \|\mathbf{v}\|^2\right).$$

It follows that any length-preserving transformation in R^3 preserves the dot product, and thus the angle between vectors. Rotations are often defined as linear transformations that preserve the inner product on R^3. This is equivalent to requiring them to preserve length.

Orthogonal and Rotation Matrices

Every rotation maps an orthonormal basis of R^3 to another orthonormal basis. Like any linear transformation of finite-dimensional vector spaces, a rotation can always be represented by a matrix. Let R be a given rotation. With respect to the standard basis (e_1, e_2, e_3) of R^3 the columns of R are given by (Re_1, Re_2, Re_3). Since the standard basis is orthonormal, the columns of R form another orthonormal basis. This orthonormality condition can be expressed in the form

$$R^T R = I$$

where R^T denotes the transpose of R and I is the 3×3 identity matrix. Matrices for which this property holds are called orthogonal matrices. The group of all 3×3 orthogonal matrices is denoted O(3), and consists of all proper and improper rotations.

In addition to preserving length, proper rotations must also preserve orientation. A matrix will preserve or reverse orientation according to whether the determinant of the matrix is positive or negative. For an orthogonal matrix R, note that $\det R^T = \det R$ implies $(\det R)^2 = 1$ so that $\det R = \pm 1$. The subgroup of orthogonal matrices with determinant +1 is called the *special orthogonal group*, denoted SO(3). Thus every rotation can be represented uniquely by an orthogonal matrix with unit determinant. Moreover, since composition of rotations corresponds to matrix multiplication, the rotation group is isomorphic to the special orthogonal group SO(3). Improper rotations correspond to orthogonal matrices with determinant −1, and they do not form a group because the product of two improper rotations is a proper rotation.

Group Structure

The rotation group is a group under function composition (or equivalently the product of linear transformations). It is a subgroup of the general linear group consisting of all invertible linear transformations of Euclidean space. Furthermore, the rotation group is nonabelian. That is, the order in which rotations are composed makes a difference.

For example, a quarter turn around the positive x-axis followed by a quarter turn around the positive y-axis is a different rotation than the one obtained by first rotating around y and then x. The orthogonal group, consisting of all proper and improper rotations, is generated by reflections. Every proper rotation is the composition of two reflections, a special case of the Cartan–Dieudonné theorem.

Chapter

7

The Hypersphere of Rotations

Visualising the Hypersphere

It is interesting to consider the space as the three dimensional sphere S^3, the boundary of a disk in 4-dimensional Euclidean space. For doing this, we will have to define how will we represent a rotation with this 4D-embedded surface. The way in which the radius can be used to specify the angle of rotation is not straightforward. It can be related to circles of latitude in a sphere with a defined north pole and is explained following.

Beginning at the north pole of a sphere in three dimensional space, we specify the point at the north pole to represent the identity rotation. In the case of the identity rotation, no axis of rotation is defined, and the angle of rotation (zero) is irrelevant. A rotation having a very small rotation angle can be specified by a slice through the sphere parallel to the xy plane and very near the north pole. The circle defined by this slice will be very small, corresponding to the small angle of the rotation. As the rotation angles become larger, the slice moves southward, and the circles become larger until the equator of the sphere is reached, which will correspond to a rotation angle of 180 degrees. Continuing southward, the radii of the circles now become smaller (corresponding to the absolute value of the angle of the rotation considered as a negative number). Finally, as the south pole is reached, the circles shrink once more to the identity rotation, which is also specified as the point at the south pole. Notice that a number of characteristics of such rotations and their representations can be seen by this visualisation.

The space of rotations is continuous, each rotation has a neighbourhood of rotations which are nearly the same, and this neighbourhood becomes flat as the neighbourhood shrinks.

Aliases

Also, each rotation is actually represented by two antipodal points on the sphere, which are at opposite ends of a line through the centre of the sphere. This reflects the fact that each rotation can be represented as a rotation about some axis, or, equivalently, as a negative rotation about an axis pointing in the opposite direction (a so-called double cover). The "latitude" of a circle representing a particular rotation angle will be half of the angle represented by that rotation, since as the point is moved from the north to south pole, the latitude ranges from zero to 180 degrees, while the angle of rotation ranges from 0 to 360 degrees. (the "longitude" of a point then represents a particular axis of rotation.) Note however that this set of rotations is not closed under composition.

Two successive rotations with axes in the *xy* plane will not necessarily give a rotation whose axis lies in the *xy* plane, and thus cannot be represented as a point on the sphere. This will not be the case with a general rotation in 3-space, which do form a closed set under composition.

This visualisation can be extended to a general rotation in 3 dimensional space. The identity rotation is a point, and a small angle of rotation about some axis can be represented as a point on a sphere with a small radius. As the angle of rotation grows, the sphere grows, until the angle of rotation reaches 180 degrees, at which point the sphere begins to shrink, becoming a point as the angle approaches 360 degrees (or zero degrees from the negative direction). This set of expanding and contracting spheres represents a hypersphere in four dimensional space (a 3-sphere).

Just as in the simpler example above, each rotation represented as a point on the hypersphere is matched by its antipodal point on that hypersphere. The "latitude" on the hypersphere will be half of the corresponding angle of rotation, and the neighbourhood of any point will become "flatter" (i.e. be represented by a 3-D Euclidean space of points) as the neighbourhood shrinks.

This behaviour is matched by the set of unit quaternions: A general quaternion represents a point in a four dimensional space, but constraining it to have unit magnitude yields a three dimensional space equivalent to the surface of a hypersphere. The magnitude of the unit quaternion will be unity, corresponding to a hypersphere of unit radius.

The vector part of a unit quaternion represents the radius of the 2-sphere corresponding to the axis of rotation, and its magnitude is the sine of half the angle of rotation. Each rotation is represented by two unit quaternions of opposite sign, and, as in the space of rotations in three dimensions, the quaternion product of two unit quaternions will yield a unit quaternion. Also, the space of unit quaternions is "flat" in any infinitesimal neighbourhood of a given unit quaternion.

Problems of the Parametrisations

There are problems in using these as more than local charts, to do with their multiple-valued nature, and singularities. That is, one must be careful above all to work only with diffeomorphisms in the definition of chart. Problems of this sort are inevitable, since SO(3) is diffeomorphic to real projective space RP^3, which is a quotient of S^3 by identifying antipodal points, and charts try to model a manifold using R^3.

This explains why, for example, the Euler angles appear to give a variable in the 3-torus, and the unit quaternions in a 3-sphere. The uniqueness of the representation by Euler angles breaks down at some points (cf. gimbal lock), while the quaternion representation is always a double cover, with q and $-q$ giving the same rotation.

If we use a skew-symmetric matrix, every 3×3 skew-symmetric matrix is determined by 3 parametres, and so at first glance, the parametre space is R^3. Exponentiating such a matrix results in an orthogonal 3×3 matrix of determinant 1 – in other words, a rotation matrix, but this is a many-to-one map.

Note that it is not a covering map – while it is a local homeomorphism near the origin, it is not a covering map at rotations by 180 degrees. It is possible to restrict these matrices to a ball around the origin in R^3 so that rotations do not exceed 180 degrees, and this will be one-to-one, except for rotations by 180 degrees, which correspond to the boundary S^2, and these identify antipodal points – this is the cut locus. The 3-ball with this identification of the boundary is RP^3. A similar situation holds for applying a Cayley transform to the skew-symmetric matrix.

Axis angle gives parametres in $S^2 \times S^1$; if we replace the unit vector by the actual axis of rotation, so that n and $-n$ give the same axis line, the set of axis becomes RP^2, the real projective plane. But since rotations around n and $-n$ are parameterised by opposite values of θ, the result is an S^1 bundle over RP^2, which turns out to be RP^3.

Fractional linear transformations use four complex parametres, a, b, c, and d, with the condition that ad-bc is non-zero. Since multiplying all four parametres by the same complex number does not change the parametre, we can insist that $ad\text{-}bc = 1$. This suggests writing (a,b,c,d) as a 2×2 complex matrix of determinant 1, that is, as an element of the special linear group SL(2,C). But not all such matrices produce rotations: conformal maps on S^2 are also included. To only get rotations we insist that d is the complex conjugate of a, and c is the negative of the complex conjugate of b. Then we have two complex numbers, a and b, subject to $|a|^2+|b|^2=1$. If we write $a + bj$, this is a quaternion of unit length.

Ultimately, since R^3 is not RP^3, there will be a problem with each of these approaches. In some cases, we need to remember that certain parametre values result in the same rotation, and to remove this issue, boundaries must be set up, but then a path through this region in R^3 must then suddenly jump to a different region when it crosses a boundary. Gimbal lock is a problem when the derivative of the map is not full rank, which occurs with Euler angles and Tait–Bryan angles, but not for the other choices. The quaternion representation has none of these problems (being a two-to-one mapping everywhere), but it has 4 parametres with a condition (unit length), which sometimes makes it harder to see the three degrees of freedom available.

Applications

One area in which these considerations, in some form, become inevitable, is the kinematics of a rigid body. One can take as definition the idea of a curve in the Euclidean group $E(3)$ of three-dimensional Euclidean space, starting at the identity (initial position).

The translation subgroup T of $E(3)$ is a normal subgroup, with quotient SO(3) if we look at the subgroup $E^+(3)$ of direct isometries only (which is reasonable in kinematics). The translational part can be decoupled from the rotational part in standard Newtonian kinematics by considering the motion of the centre of mass, and rotations of the rigid body about the centre of mass. Therefore any rigid body movement leads directly to SO(3), when we factor out the translational part.

These identifications illustrate that SO(3) is connected but not simply connected. As to the latter, in the ball with antipodal surface points identified, consider the path running from the "north pole" straight through the centre down to the south pole. This is a closed loop, since the north pole and the south pole are identified. This loop cannot be shrunk to a point, since no matter how you deform the loop, the start and end point have to remain antipodal, or else the loop will "break open". In terms of rotations, this loop represents a continuous sequence of rotations about the z-axis starting and ending at the identity rotation (i.e. a series of rotation through an angle φ where φ runs from 0 to 2π).

Surprisingly, if you run through the path twice, i.e., from north pole down to south pole and back to the north pole so that φ runs from 0 to 4π, you get a closed loop which *can* be shrunk to a single point: first move the paths continuously to the ball's surface, still connecting north pole to south pole twice. The second half of the path can then be mirrored over to the antipodal side without changing the path at all. Now we have an ordinary closed loop on the surface of the ball, connecting the north pole to itself along a great circle. This circle can be shrunk to the north pole without problems.

The same argument can be performed in general, and it shows that the fundamental group of SO(3) is cyclic group of order 2. In physics applications, the non-triviality of the fundamental group allows for the existence of objects known as spinors, and is an important tool in the development of the spin-statistics theorem.

The universal cover of SO(3) is a Lie group called Spin(3). The group Spin(3) is isomorphic to the special unitary group SU(2); it is also diffeomorphic to the unit 3-sphere S^3 and can be understood as the group of unit quaternions (i.e. those with absolute value 1). The connection between quaternions and rotations, commonly exploited in computer graphics, is explained in quaternions and spatial rotations. The map from S^3 onto SO(3) that identifies antipodal points of S^3 is a surjective homomorphism of Lie groups, with kernel $\{\pm 1\}$. Topologically, this map is a two-to-one covering map.

Rotation Formalisms in Three Dimensions

In geometry, various formalisms exist to express a rotation in three dimensions as a mathematical transformation. In physics, this concept extends to classical mechanics where rotational (or angular) kinematics is the science of describing with numbers the purely rotational motion of an object.

According to Euler's rotation theorem the general displacement of a rigid body (or three-dimensional coordinate system) with one point fixed is described by a rotation about some axis. This allows the use of rotations to express orientations as a single rotation from a reference placement in space of the rigid body (or coordinate system). Furthermore, such a rotation may be uniquely described by a minimum of three parametres. However, for various reasons, there are several ways to represent it. Many of these representations use more than the necessary minimum of three parametres, although each of them still has only three degrees of freedom.

An example where rotation representation is used is in computer vision, where an automated observer needs to track a target. Let's consider a rigid body, with an orthogonal right-handed triad $\hat{u}$, $\hat{v}$, and $\hat{w}$ of unit vectors fixed to its body (representing the three axes of the object's coordinate system). The basic problem is to specify the orientation of this triad, and hence the rigid body, in terms of the reference coordinate system (in our case the observer's coordinate system).

Rotation Matrix

The above mentioned triad of unit vectors is also called a basis. Specifying the coordinates (scalar components) of this basis in its current (rotated) position, in terms of the reference (non-rotated) coordinate axes, will completely describe the rotation. The three unit vectors $\hat{u}, \hat{v}$ and $\hat{w}$ which form the rotated basis each consist of 3 coordinates, yielding a total of 9 parametres. These parametres can be written as the elements of a 3 × 3 matrix A, called a rotation matrix. Typically, the coordinates of each of these vectors are arranged along a column of the matrix (however, beware that an alternative definition of rotation matrix exists and is widely used, where the vectors coordinates defined above are arranged by rows)

$$\mathbf{A} = \begin{bmatrix} \hat{\mathbf{u}}_x & \hat{\mathbf{v}}_x & \hat{\mathbf{w}}_x \\ \hat{\mathbf{u}}_y & \hat{\mathbf{v}}_y & \hat{\mathbf{w}}_y \\ \hat{\mathbf{u}}_z & \hat{\mathbf{v}}_z & \hat{\mathbf{w}}_z \end{bmatrix}$$

The elements of the rotation matrix are not all independent – as Euler's rotation theorem dictates, the rotation matrix has only three degrees of freedom. The rotation matrix has the following properties:

- A is a real, orthogonal matrix, hence each of its rows or columns represents a unit vector.

- The eigenvalues of A are $\{1, e^{\pm i\theta}\} = \{1, \cos(\theta) + i\sin(\theta), \cos(\theta) - i\sin(\theta)\}$ where i is the standard imaginary unit with the property $i^2 = -1$
- The determinant of A is +1, equivalent to the product of its eigenvalues.
- The trace of A is $1+2\cos(\theta)$, equivalent to the sum of its eigenvalues.

The angle θ which appears in the eigenvalue expression corresponds to the angle of the Euler axis and angle representation. The eigenvector corresponding with the eigenvalue of 1 is the accompanying Euler axis, since the axis is the only (nonzero) vector which remains unchanged by left-multiplying (rotating) it with the rotation matrix.

The above properties are equivalent to:

$$
\begin{aligned}
|\hat{\mathbf{u}}| = |\hat{\mathbf{v}}| &= 1 \\
\hat{\mathbf{u}} \cdot \hat{\mathbf{v}} &= 0 \\
\hat{\mathbf{u}} \times \hat{\mathbf{v}} &= \hat{\mathbf{w}}
\end{aligned}
$$

which is another way of stating that $\hat{\mathbf{u}}, \hat{\mathbf{v}}, \hat{\mathbf{w}}$ form a 3D orthonormal basis. Note that the statements above constitute a total of 6 conditions (the cross product contains 3), leaving the rotation matrix with just 3 degrees of freedom as required. Two successive rotations represented by matrices A_1 and A_2 are easily combined as follows: $\mathbf{A}_{\text{total}} = \mathbf{A}_2\mathbf{A}_1$ (Note the order, since the vector being rotated is multiplied from the right). The ease by which vectors can be rotated using a rotation matrix, as well as the ease of combining successive rotations, make the rotation matrix a very useful and popular way to represent rotations, even though it is less concise than other representations.

Axis-angle Representation

The axis-angle representation of a rotation, also known as the exponential coordinates of a rotation, parametreises a rotation by two values: a unit vector indicating the direction of a directed axis (straight line), and an angle describing the magnitude of the rotation about the axis. The rotation occurs in the sense prescribed by the right-hand rule.

This representation evolves from Euler's rotation theorem, which implies that any rotation or sequence of rotations of a rigid body in a three-dimensional space is equivalent to a pure rotation about a single fixed axis.

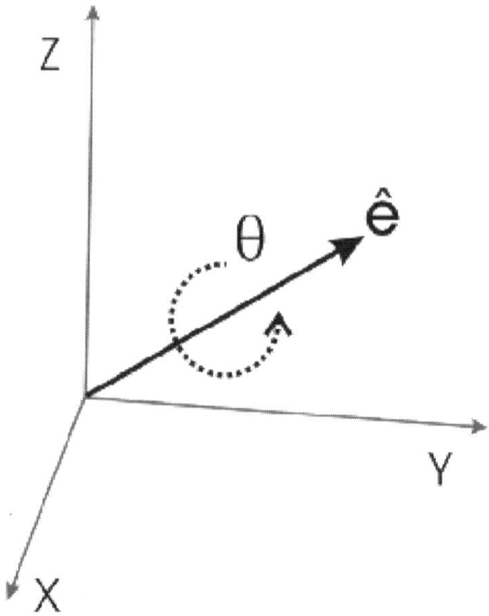

Figure: *A visualisation of a rotation represented by an Euler axis and angle.*

The axis-angle representation is equivalent to the more concise rotation vector, or Euler vector representation. In this case, both the axis and the angle are represented by a non-normalised vector codirectional with the axis whose magnitude is the rotation angle.

Rodrigues' rotation formula can be used to apply to a vector a rotation represented by an axis and an angle.

Uses

The axis-angle representation is convenient when dealing with rigid body dynamics. It is useful to both characterise rotations, and also for converting between different representations of rigid body motion, such as homogeneous transformations and twists.

Example

Say you are standing on the ground and you pick the direction of gravity to be the negative z direction. Then if you turn to your left, you will travel $\frac{\pi}{2}$ radians (or 90 degrees) about the z axis. In axis-angle representation, this would be

$$\langle \text{axis}, \text{angle} \rangle = \left(\begin{bmatrix} a_x \\ a_y \\ a_z \end{bmatrix}, \theta \right) = \left(\begin{bmatrix} 0 \\ 0 \\ 1 \end{bmatrix}, \frac{\pi}{2} \right)$$

Rotation Vector

The above example can be represented as a rotation vector with a magnitude of $\frac{\pi}{2}$ pointing in the z direction.

$$\begin{bmatrix} 0 \\ 0 \\ \frac{\pi}{2} \end{bmatrix}$$

This is the product of the angle and vector. It is more compact and is used for the exponential and log maps involving this representation.

Simultaneous Orthogonal Rotation Angle

Simultaneous Orthogonal Rotations Angle (SORA) is a vector representing angular orientation of a rigid-body relative to some reference frame. The components of this vector are equal to the angles of three simultaneous rotations around the rigid body's intrinsic coordinate system axes, initially aligned with the axes of the reference frame, needed to move the rigid body to its current angular orientation.

Every angular orientation can be represented by a single rotation. Therefore, the three simultaneous orthogonal rotations actually represent a single rotation around a certain axis v and for a certain angle φ. The orientation and magnitude of SORA are equal to this equivalent single rotation axis v and angle φ, respectively. SORA is therefore the rotation vector. Denoting the angles of the three simultaneous orthogonal rotations with φ_x, φ_y, and φ_z, SORA is equal to:

$$\mathbf{\Phi} = \begin{bmatrix} \varphi_x \\ \varphi_y \\ \varphi_z \end{bmatrix} = \mathbf{v}\phi$$

where v and are the equivalent single rotation axis and angle respectively, equal to:

$$\mathbf{v} = \frac{1}{\sqrt{\varphi_x^2 + \varphi_y^2 + \varphi_z^2}} \begin{bmatrix} \varphi_x \\ \varphi_y \\ \varphi_z \end{bmatrix},$$

$$\phi = \sqrt{\varphi_x^2 + \varphi_y^2 + \varphi_z^2}.$$

An analytic derivation of these single rotation equivalent axis and angle is achieved by decomposing the total rotation to an infinite sequence of infinitesimally small rotations.

The most common scenario, where simultaneous rotation around three orthogonal axes can be observed is when performing measurements with a gyroscope.

The three simultaneous orthogonal rotations measured with a 3D gyroscope represent a single rotation around a certain axis for a certain angle.

When the measured angular velocities or their proportions are constant, this rotation and the resulting angular orientation can be uniquely represented with SORA. When this condition is met, the orientation of the rotation axis does not change and the components of SORA are equal to the angles of three simultaneous rotations measured with a 3D gyroscope.

Using gyroscope measurements that provide rotation angle estimation, it is reasonable to combine angular velocities around three orthogonal axes in vector form:

$$\mathbf{\Omega} = \begin{bmatrix} \omega_x \\ \omega_y \\ \omega_z \end{bmatrix}.$$

Angular velocity vector Ω is orthogonal to the rotation plane, and its magnitude is equal to the angular velocity in that plane. When the measured angular velocities or their proportions are approximately constant, the orientation of the rotation axis is also constant. It then holds:

$$\mathbf{\Omega} = \frac{\mathbf{\Phi}}{T}.$$

SORA represents a unique angular orientation of a rigid-body relative to some reference frame. Using SORA as a generalised representation of angular orientation, we can avoid the problem of rotation non-commutativity. As SORA indicates a single rotation for a specific angular orientation, the Gimbal problem is here avoided. Because of the mentioned, SORA is much more suitable for representation of angular orientation than widely used Euler angles.

SORA is simple and well-suited for use in the real-time calculation of orientation based on angular velocity measurements derived using a gyroscope, as long as the measured angular velocities or their proportions are approximately constant. Under this condition, using SORA, orientation calculation only requires a single step. SORA is appropriate for use in a number of applications that require angular orientation information, including navigation and motion tracking.

Finally, SORA helps interpret the meaning of the rotation vector with measurable quantities.

Rotating a Vector

Rodrigues' rotation formula (named after Olinde Rodrigues) is an efficient algorithm for rotating a vector in space, given a rotation axis and an angle of rotation. In other words, the Rodrigues formula provides an algorithm to compute the exponential map from so(3) to SO(3) without computing the full matrix exponent (the rotation matrix).

If v is a vector in $\mathbb{R}^3$ and ω is a unit vector describing an axis of rotation about which we want to rotate v by an angle θ (in a right-handed sense), the Rodrigues formula to obtain the rotated vector is:

$$\mathbf{v}_{rot} = \mathbf{v}\cos\theta + (\boldsymbol{\omega}\times\mathbf{v})\sin\theta + \boldsymbol{\omega}(\boldsymbol{\omega}\cdot\mathbf{v})(1-\cos\theta).$$

This is more efficient than converting ω and θ into a rotation matrix, and using the rotation matrix to compute the rotated vector.

Relationship to Other Representations

There are many ways to represent a rotation. It is useful to understand how different representations relate to one another, and how to convert between them.

Euler Angles

The Euler angles are three angles introduced by Leonhard Euler to describe the orientation of a rigid body. To describe such an orientation in 3-dimensional Euclidean space three parametres are required. They can be given in several ways, Euler angles being one of them.

Euler angles also represent three composed rotations that move a reference frame to a given referred frame. This is equivalent to saying that any orientation can be achieved by composing three elemental rotations (rotations around a single axis), and also equivalent to saying that any rotation matrix can be decomposed as a product of three elemental rotation matrices. Without considering the possibilities of different signs for the angles or moving the reference frame, there are twelve different conventions divided in two groups. One of them is called "proper" Euler angles and the other Tait–Bryan angles. Sometimes "Euler angles" is used for all of them.

Proper Euler Angles

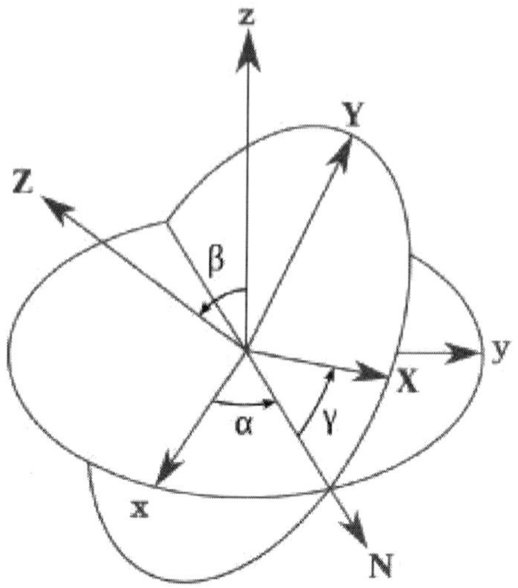

Figure: *Euler angles – The xyz (fixed) system is shown in blue, the XYZ (rotated) system is shown in red. The line of nodes, labeled N, is shown in green.*

Sometimes Tait-Bryan angles are called Euler angles. When this is done, it is necessary to specify when classic Euler angles are referred, using an adjective like "proper" or similar.

Classic Definition

Euler angles are a means of representing the spatial orientation of any frame (coordinate system) as a composition of rotations from a frame of reference (coordinate system). In the following, the fixed system is denoted in lower case (x,y,z) and the rotated system is denoted in upper case letters (X,Y,Z).

The geometrical definition (referred sometimes as static) is based on a reference frame and one whose orientation we want to describe, first we define the line of nodes (N) as the intersection of the xy and the XY coordinate planes (in other words, line of nodes is the line perpendicular to both z and Z axis). Then we define its Euler angles as:

- α (or φ) is the angle between the x-axis and the line of nodes.
- β (or θ) is the angle between the z-axis and the Z-axis.
- γ (or ψ) is the angle between the line of nodes and the X-axis.

Euler angles between two frames are defined only if both frames have the same handedness. Euler angles are just one of the several ways of specifying the relative orientation of two such coordinate systems. Different authors may use different sets of angles to describe these orientations, or different names for the same angles, leading to different conventions. Therefore any discussion employing Euler angles should always be preceded by their definition.

Unless otherwise stated, this article will use the convention described in the adjacent drawing, usually named Z-X'-Z".

Angle Signs and Ranges

Normally, angles are defined in such a way that they are positive when they rotate counter-clockwise (the rotation depending on which side of the rotation plane they are observed from, with the positive side being the one containing the positive axis of rotation).

About the ranges:

- α and γ range are defined modulo 2π radians. A valid range could be $(-\pi, \pi]$.
- β range covers π radians (but can't be said to be modulo π). For example could be $[0, \pi]$ or $[-\pi/2, \pi/2]$.

The angles α, β and γ are uniquely determined except for the singular case that the xy and the XY planes are identical, the z axis and the Z axis having the same or opposite directions. Indeed, if the z-axis and the Z-axis are the same, $\beta = 0$ and only $(\alpha + \gamma)$ is uniquely defined (not the individual values), and, similarly, if the z-axis and the Z-axis are opposite, $\beta = \pi$ and only $(\alpha - \gamma)$ is uniquely defined (not the individual values). These ambiguities are known as gimbal lock in applications.

Conventions for Proper Euler Angles

The geometric definition for Euler angles is based in the line of nodes, which must be the intersection of two homologous planes (planes overlapping when the angles are zero). There are six possibilities of choosing the proper Euler angles. Using the static definition they correspond to the three possible homogeneous combinations of planes (*XY*, *XZ* and *YZ*) with the two possible options to measure the angles from (given the line of nodes by the *XY* planes for example, it can be taken *X-N* or *Y-N* as first angle). Hence the six possibilities.

Though is possible to define the angles by geometry, normally the intrinsic rotations equivalence or the extrinsic rotations equivalence are used instead. According with these equivalences, proper Euler angles are equivalent to three combined rotations repeating exactly one axis. The six possible conventions are named after their equivalent rotation composition. When the intrinsic rotations equivalence is used to name the possible conventions of Euler Angles the names are like Z-X'-Z". This means that they are equivalent to three concatenated intrinsic rotations around some moving axes Z, X' and Z" in that order. This composition is non-commutative. It has to be applied in such a way that in the beginning one of the intrinsic axis moves together with the line of nodes. The above diagram convention is usually named this way.

When the extrinsic rotations equivalence is used the given angles are inverted respect the former, meaning that the first angle is the intrinsic rotation and the last one the precession.

If a loose naming convention, like *Z-X-Z* is used, the name of both conventions would be indistinguishable, even if the angles' order is the opposite. To specify that the given order means intrinsic composition, sometimes a similar notation is used, but stating explicitly which rotation axis are different for each step, as in *Z-X'-Z"*. Using this notation, *Z-X-Z* would mean extrinsic composition. It is also possible to use alternative names like *x-y-x* or *X-Y'-X"* to clarify.

No geometrical information is lost when using the rotation equivalence because the static parametres can be calculated from the name of the convention. For example, given the convention *X-Y'-Z"*, the first rotation is perpendicular to "*x*" and the third one to "*Z*". Therefore the planes are the *yz* and the *XY*, and the line of nodes is the intersection of these two.

Tait–Bryan Angles

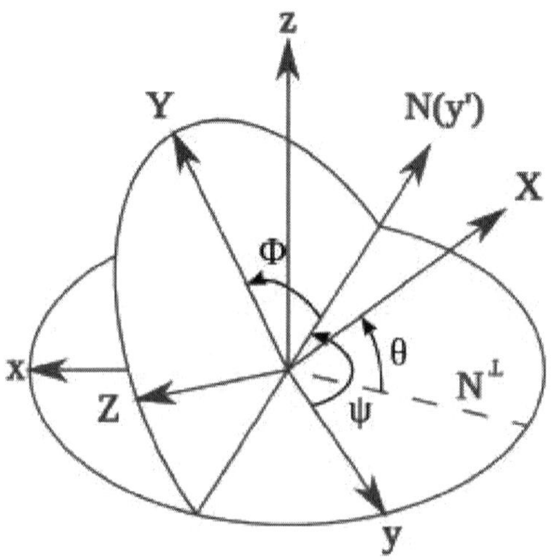

Figure: *Tait–Bryan angles. ZYX (intrinsic composition) convention*

The second type of conventions is called Tait–Bryan angles, after Peter Guthrie Tait and George H. Bryan, also known as Nautical or Cardan angles, after Cardan.

Their static difference is the definition for the line of nodes. In the first case two homologous planes were used. In the second one, they are replaced by non-homologous planes (perpendicular when angles are zero).

While proper Euler angles are equivalent to three combined rotations repeating exactly one axis, Tait–Bryan angles are equivalent to three composed rotations in different axes.

There are also six Tait–Bryan combinations. They come from the two possible non-homogeneous planes that exist when one is given (given *XY*, there are two non-homogeneous, *XZ* and *YZ*). The three possible planes at the reference frame multiplied by the two options for each one yield the six possible conventions.

There are six possible combinations of this kind, and all of them behave in an identical way. Using the intrinsic rotations equivalence, Tait–Bryan angles correspond with the three rotations with a different axis. *Z–X–Y* for example. There are also six possibilities of this kind. The enclosed image shows the *ZXY* convention. The other five proper conventions are obtained by selecting different axes of rotation.

These three angles are normally called "Heading, Elevation and Bank", or "Yaw, Pitch and Roll". The second terms have to be used carefully because they are also the names for the three aircraft principal axes.

Double Naming Problem

For either Euler or Tait-Bryan angles, it is very simple to convert from an intrinsic (rotating axes) to an extrinsic (static axes) convention, and vice-versa: just swap the order of the operations. An (α, β, γ) rotation using *X-Y-Z* intrinsic convention is equivalent to a (γ, β, α) rotation using *Z-Y-X* extrinsic convention; this is true for all Euler or Tait-Bryan axis combinations.

For Tait-Bryan angles, also intrinsic and extrinsic conventions can be used, giving therefore two meanings for every convention name. For example, *X-Y-Z*, using intrinsic convention, means that a *X*-rotation is performed, composing intrinsic rotations *Y* and *Z* later, but using extrinsic convention means that after the *X* rotation, extrinsic rotations *Y* and *Z* are performed. The meaning is different in both cases. Alternative naming schemas like *z-x-y* versus *Z-X'-Y"* could be used to prevent confusion.

Relationship with Physical Motions

Euler angles can be considered as the result of three composed rotations and these conventions are named according to this composition. Since intrinsic rotations produce the same result as extrinsic rotations inverted, there are two possible names for any static convention. For example the intrinsic *ZXY* and the extrinsic *yxz* have the same static parametres.

Euler Angles as Composition of Intrinsic Rotations

Starting with an initial set of mobile axes, say *XYZ* overlapping the reference axes *xyz*, a composition of three intrinsic rotations (rotations only about the mobile frame axes, assuming active composition) can be used to reach any target frame with an origin coincident with that of *XYZ* from the reference frame. The value of the rotations are the Euler Angles.

The position of the mobile axes can be reached using three rotations with angles α, β, γ in three ways *equivalent* to the former definition, as follows:

The *XYZ* system rotates while the *xyz* is fixed. Starting with the *XYZ* system overlapping the reference frame *xyz*, the same rotations as before can be performed using only rotations around the mobile axes XYZ.

- Rotate the *XYZ*-system about the *Z*-axis by α. The *X*-axis now lies on the line of nodes.
- Rotate the *XYZ*-system again about the now rotated *X*-axis by β. The *Z*-axis is now in its final orientation, and the *x*-axis remains on the line of nodes.
- Rotate the *XYZ*-system a third time about the new *Z*-axis by γ.

Any convention for proper Euler angles is equivalent to three such rotations, with one axis being repeated (ZXZ for example). Tait-Bryan angles are also equivalent to three composed rotations, but in this case, all three rotations are around different axes (ZXY). Usually conventions are named according with this equivalence.

Euler Rotations

Euler rotations are defined as the movement obtained by changing one of the Euler angles while leaving the other two constant. Euler rotations are never expressed in terms of the external frame, or in terms of the co-moving rotated body frame, but in a mixture. They constitute a mixed axes of rotation system, where the first angle moves the line of nodes around the external axis z, the second rotates around the line of nodes and the third one is an intrinsic rotation around an axis fixed in the body that moves. These rotations are called Precession, Nutation, and intrinsic rotation. While they are rotations when they are applied over individual frames, only precession is valid as a rotation operator, and only precession can be expressed in general as a matrix in the basis of the space.

Gimbal Analogy

If we suppose a set of frames, able to move each with respect to the former according to just one angle, like a gimbal, there will exist an external fixed frame, one final frame and two frames in the middle, which are called "intermediate frames". The two in the middle work as two gimbal rings that allow the last frame to reach any orientation in space.

Intermediate Frames

The gimbal rings indicate some intermediate frames. They can be defined statically too. Taking some vectors i, j and k over the axes x, y and z, and vectors I, J, K over X, Y and Z, and a vector N over the line of nodes, some intermediate frames can be defined using the vector cross product, as following:

- origin: $[i,j,k]$ (where $k = i \times j$)
- first: $[N,k \times N,k]$
- second: $[N,K \times N,K]$

·final: $[I,J,K]$

These intermediate frames are equivalent to those of the gimbal. They are such that they differ from the previous one in just a single elemental rotation. This proves that:

- Any target frame can be reached from the reference frame just composing three rotations.
- The values of these three rotations are exactly the Euler angles of the target frame.

Properties

The Euler angles form a chart on all of SO(3), the special orthogonal group of rotations in 3D space. The chart is smooth except for a polar coordinate style singularity along $\beta = 0$.

The space of rotations is called in general "The Hypersphere of rotations", though this is a misnomer: the group Spin(3) is isometric to the hypersphere S^3, but the rotation space SO(3) is instead isometric to the real projective space RP^3 which is a 2-fold quotient space of the hypersphere. This 2-to-1 ambiguity is the mathematical origin of spin in physics.

A similar three angle decomposition applies to SU(2), the special unitary group of rotations in complex 2D space, with the difference that β ranges from 0 to 2π. These are also called Euler angles.

The Haar measure for Euler angles has the simple form sin(β).dα.dβ.dγ, usually normalised by a factor of $1/8\pi^2$. For example, to generate uniformly randomised orientations, let α and γ be uniform from 0 to 2π, let z be uniform from −1 to 1, and let β = arccos(z).

Higher Dimensions

It is possible to define parametres analogous to the Euler angles in dimensions higher than three.

The number of degrees of freedom of a rotation matrix is always less than the dimension of the matrix squared. That is, the elements of a rotation matrix are not all completely independent. For example, the rotation matrix in dimension 2 has only one degree of freedom, since all four of its elements depend on a single angle of rotation. A rotation matrix in dimension 3 (which has nine elements) has three degrees of freedom, corresponding to each independent rotation, for example by its three Euler angles or a magnitude one (unit) quaternion.

In SO(4) the rotation matrix is defined by two quaternions, and is therefore 6-parametric (three degrees of freedom for every quaternion). The 4x4 rotation matrices have therefore 6 out of 16 independent components.

Any set of 6 parametres that define the rotation matrix could be considered an extension of Euler angles to dimension 4.

In general, the number of euler angles in dimension D is quadratic in D; since any one rotation consists of choosing two dimensions to rotate between, the total number of rotations available in dimension D is $N_{rot} = \binom{D}{2} = D(D-1)/2$, which for D = 2,3,4 yields $N_{rot} = 1,3,6$.

Others

Euler angles, normally in the Tait–Bryan convention, are also used in robotics for speaking about the degrees of freedom of a wrist. They are also used in Electronic stability control in a similar way.

Gun fire control systems require corrections to gun-order angles (bearing and elevation) to compensate for deck tilt (pitch and roll). In traditional systems, a stabilising gyroscope with a vertical spin axis corrects for deck tilt, and stabilises the optical sights and radar antenna. However, gun barrels point in a direction different from the line of sight to the target, to anticipate target movement and fall of the projectile due to gravity, among other factors. Gun mounts roll and pitch with the deck plane, but also require stabilisation. Gun orders include angles computed from the vertical gyro data, and those computations involve Euler angles.

Euler angles are also used extensively in the quantum mechanics of angular momentum. In quantum mechanics, explicit descriptions of the representations of *SO(3)* are very important for calculations, and almost all the work has been done using Euler angles. In the early history of quantum mechanics, when physicists and chemists had a sharply negative reaction towards abstract group theoretic methods (called the *Gruppenpest*), reliance on Euler angles was also essential for basic theoretical work. In materials science, crystallographic texture (or preferred orientation) can be described using Euler angles. In texture analysis, the Euler angles provide the necessary mathematical depiction of the orientation of individual crystallites within a polycrystalline material, allowing for the quantitative description of the macroscopic material. The most common definition of the angles is due to Bunge and corresponds to the *ZXZ* convention. It is important to note, however, that the application generally involves axis transformations of tensor quantities, i.e. passive rotations. Thus the matrix that corresponds to the Bunge Euler angles is the transpose of that shown in the table above.

Many mobile computing devices contain accelerometres which can determine these devices' Euler angles with respect to the earth's gravitational attraction. These are used in applications such as games, bubble level simulations, and kaleidoscopes.

Quaternions and Spatial Rotation

Unit quaternions provide a convenient mathematical notation for representing orientations and rotations of objects in three dimensions. Compared to Euler angles they are simpler to compose and avoid the problem of gimbal lock. Compared to rotation matrices they are more numerically stable and may be more efficient. Quaternions have found their way into applications in computer graphics, computer vision, robotics, navigation, molecular dynamics, flight dynamics., and orbital mechanics of satellites.

When used to represent rotation, unit quaternions are also called versors, or rotation quaternions. When used to represent an orientation (rotation relative to a reference position), they are called orientation quaternions or attitude quaternions.

Quaternion Rotation Operations

A very formal explanation of the properties used in this section is given by Altmann.

The Hypersphere of Rotations

Visualising the Space of Rotations: Unit quaternions represent the mathematical space of rotations in three dimensions in a very straightforward way. The correspondence between rotations and quaternions can be understood by first visualising the space of rotations itself.

In order to visualise the space of rotations, it helps to consider a simpler case. Any rotation in three dimensions can be described by a rotation by some angle about some axis. Consider the special case in which the axis of rotation lies in the xy plane.

We can then specify the axis of one of these rotations by a point on a circle, and we can use the radius of the circle to specify the angle of rotation. Similarly, a rotation whose axis of rotation lies in the xy plane can be described as a point on a sphere of fixed radius in three dimensions.

Beginning at the north pole of a sphere in three dimensional space, we specify the point at the north pole to be the identity rotation (a zero angle rotation). Just as in the case of the identity rotation, no axis of rotation is defined, and the angle of rotation (zero) is irrelevant.

A rotation having a very small rotation angle can be specified by a slice through the sphere parallel to the xy plane and very near the north pole. The circle defined by this slice will be very small, corresponding to the small angle of the rotation. As the rotation angles become larger, the slice moves in the negative z direction, and the circles become larger until the equator of the sphere is reached, which will correspond to a rotation angle of 180 degrees. Continuing southward, the radii of the circles now become smaller (corresponding to the absolute value of the angle of the rotation considered as a negative number). Finally, as the south pole is reached, the circles shrink once more to the identity rotation, which is also specified as the point at the south pole.

Notice that a number of characteristics of such rotations and their representations can be seen by this visualisation. The space of rotations is continuous, each rotation has a neighbourhood of rotations which are nearly the same, and this neighbourhood becomes flat as the neighbourhood shrinks. Also, each rotation is actually represented by two antipodal points on the sphere, which are at opposite ends of a line through the centre of the sphere. This reflects the fact that each rotation can be represented as a rotation about some axis, or, equivalently, as a negative rotation about an axis pointing in the opposite direction (a so-called double cover).

The "latitude" of a circle representing a particular rotation angle will be half of the angle represented by that rotation, since as the point is moved from the north to south pole, the latitude ranges from zero to 180 degrees, while the angle of rotation ranges from 0 to 360 degrees. (the "longitude" of a point then represents a particular axis of rotation.) Note however that this set of rotations is not closed under composition. Two successive rotations with axes in the xy plane will not necessarily give a rotation whose axis lies in the xy plane, and thus cannot be represented as a point on the sphere. This will not be the case with a general rotation in 3-space, in which rotations do form a closed set under composition. This visualisation can be extended to a general rotation in 3 dimensional space. The identity rotation is a point, and a small angle of rotation about some axis can be represented as a point on a sphere with a small radius.

As the angle of rotation grows, the sphere grows, until the angle of rotation reaches 180 degrees, at which point the sphere begins to shrink, becoming a point as the angle approaches 360 degrees (or zero degrees from the negative direction).

This set of expanding and contracting spheres represents a hypersphere in four dimensional space (a 3-sphere). Just as in the simpler example above, each rotation represented as a point on the hypersphere is matched by its antipodal point on that hypersphere. The "latitude" on the hypersphere will be half of the corresponding angle of rotation, and the neighbourhood of any point will become "flatter" (i.e. be represented by a 3-D Euclidean space of points) as the neighbourhood shrinks. This behaviour is matched by the set of unit quaternions: A general quaternion represents a point in a four dimensional space, but constraining it to have unit magnitude yields a three dimensional space equivalent to the surface of a hypersphere. The magnitude of the unit quaternion will be unity, corresponding to a hypersphere of unit radius.

The vector part of a unit quaternion represents the radius of the 2-sphere corresponding to the axis of rotation, and its magnitude is the cosine of half the angle of rotation. Each rotation is represented by two unit quaternions of opposite sign, and, as in the space of rotations in three dimensions, the quaternion product of two unit quaternions will yield a unit quaternion. Also, the space of unit quaternions is "flat" in any infinitesimal neighbourhood of a given unit quaternion.

Parameterising the Space of Rotations

We can parameterise the surface of a sphere with two coordinates, such as latitude and longitude. But latitude and longitude are ill-behaved (degenerate) at the north and south poles, though the poles are not intrinsically different from any other points on the sphere. At the poles (latitudes +90° and −90°), the longitude becomes meaningless. It can be shown that no two-parametre coordinate system can avoid such degeneracy. We can avoid such problems by embedding the sphere in three-dimensional space and parametreising it with three Cartesian coordinates (w, x, y), placing the north pole at $(w, x, y) = (1, 0, 0)$, the south pole at $(w, x, y) = (-1, 0, 0)$, and the equator at $w = 0$, $x^2 + y^2 = 1$. Points on the sphere satisfy the constraint $w^2 + x^2 + y^2 = 1$, so we still have just two degrees of freedom though there are three coordinates. A point (w, x, y) on the sphere represents a rotation in the ordinary space around the horizontal axis directed by the vector $(x,y,0)$ by an angle $\alpha = 2\cos^{-1} w = 2\sin^{-1}\sqrt{x^2+y^2}$

In the same way the hyperspherical space of 3D rotations can be parameterised by three angles (Euler angles), but any such parameterisation is degenerate at some points on the hypersphere, leading to the problem of gimbal lock.

We can avoid this by using four Euclidean coordinates w, x, y, z, with $w^2 + x^2 + y^2 + z^2 = 1$. The point (w, x, y, z) represents a rotation around the axis directed by the vector (x,y,z) by an angle $\alpha = 2\cos^{-1} w = 2\sin^{-1}\sqrt{x^2+y^2+z^2}$.

Explaining Quaternions' Properties with Rotations

Non-commutativity

The multiplication of quaternions is non-commutative. Since the multiplication of unit quaternions corresponds to the composition of three dimensional rotations, this property can be made intuitive by showing that three dimensional rotations are not commutative in general.

Set two books next to each other. Rotate one of them 90 degrees clockwise around the z axis, then flip it 180 degrees around the x axis. Take the other book, flip it 180 around x axis first, and 90 clockwise around z later.

The two books do not end up parallel. This shows that, in general, the composition of two different rotations around two distinct spatial axes will not commute.

Are Quaternions Handed?

Note that quaternions, like the rotations or other linear transforms, are not "handed" (as in left-handed vs right-handed). Handedness of a coordinate system comes from the interpretation of the numbers in physical space. No matter what the handedness convention, rotating the X vector 90 degrees around the Z vector will yield the Y vector — the mathematics and numbers are the same.

Alternatively, if the quaternion or direction cosine matrix is interpreted as a rotation from one frame to another, then it must be either left-handed or right-handed. For example, in the above example rotating the X vector 90 degrees around the Z vector yields the Y vector only if you use the right hand rule.

If you use a left hand rule, the result would be along the negative Y vector. Transformations from quaternion to direction cosine matrix often do not specify whether the input quaternion should be left handed or right handed.

It is possible to determine the handedness of the algorithm by constructing a simple quaternion from a vector and an angle and assuming right handedness to begin with. For example, $(1+\mathbf{k})/\sqrt{2}$ has the axis of rotation along the z-axis, and a rotation angle of 90 degrees. Pass this quaternion into the quaternion to matrix algorithm. If the end result is as shown below and you wish to interpret the matrix as right-handed, then the algorithm is expecting a right-handed quaternion. If the end result is the transpose and you still want to interpret the result as a right-handed matrix, then you must feed the algorithm left-handed quaternions. To convert between left and right-handed quaternions simply negate the vector part of the quaternion.

$$\begin{pmatrix} 0 & 1 & 0 \\ -1 & 0 & 0 \\ 0 & 0 & 1 \end{pmatrix}$$

Quaternions and Other Representations of Rotations

Qualitative Description of the Advantages of Quaternions

The representation of a rotation as a quaternion (4 numbers) is more compact than the representation as an orthogonal matrix (9 numbers). Furthermore, for a given axis and angle, one can easily construct the corresponding quaternion, and conversely, for a given quaternion one can easily read off the axis and the angle. Both of these are much harder with matrices or Euler angles. In video games and other applications, one is often interested in "smooth rotations", meaning that the scene should slowly rotate and not in a single step. This can be accomplished by choosing a curve such as the spherical linear interpolation in the quaternions, with one endpoint being the identity transformation 1 (or some other initial rotation) and the other being the intended final rotation. This is more problematic with other representations of rotations. When composing several rotations on a computer, rounding errors necessarily accumulate. A quaternion that's slightly off still represents a rotation after being normalised— a matrix that's slightly off may not be orthogonal anymore and is harder to convert back to a proper orthogonal matrix. Quaternions also avoid a phenomenon called gimbal lock which can result when, for example in pitch/yaw/roll rotational systems, the pitch is rotated 90° up or down, so that yaw and roll then correspond to the same motion, and a degree of freedom of rotation is lost. In a gimbal-based aerospace inertial navigation system, for instance, this could have disastrous results if the aircraft is in a steep dive or ascent.

Conversion to and from the Matrix Representation

The orthogonal matrix corresponding to a rotation by the unit quaternion $z = a + bi + cj + dk$ (with $|z| = 1$) is given by

$$\begin{pmatrix} a^2+b^2-c^2-d^2 & 2bc-2ad & 2bd+2ac \\ 2bc+2ad & a^2-b^2+c^2-d^2 & 2cd-2ab \\ 2bd-2ac & 2cd+2ab & a^2-b^2-c^2+d^2 \end{pmatrix}.$$

From an Orthogonal Matrix to a Quaternion: One must be careful when converting a rotation matrix to a quaternion, as several straightforward methods tend to be unstable when the trace (sum of the diagonal elements) of the rotation matrix is zero or very small. For a stable method of converting an orthogonal matrix to a quaternion.

Fitting Quaternions: The above section described how to recover a quaternion q from a 3×3 rotation matrix Q.

Suppose, however, that we have some matrix Q that is not a pure rotation — due to round-off errors, for example — and we wish to find the quaternion q that most accurately represents Q. In that case we construct a symmetric 4×4 matrix

$$K = \frac{1}{3}\begin{bmatrix} Q_{xx}-Q_{yy}-Q_{zz} & Q_{yx}+Q_{xy} & Q_{zx}+Q_{xz} & Q_{yz}-Q_{zy} \\ Q_{yx}+Q_{xy} & Q_{yy}-Q_{xx}-Q_{zz} & Q_{zy}+Q_{yz} & Q_{zx}-Q_{xz} \\ Q_{zx}+Q_{xz} & Q_{zy}+Q_{yz} & Q_{zz}-Q_{xx}-Q_{yy} & Q_{xy}-Q_{yx} \\ Q_{yz}-Q_{zy} & Q_{zx}-Q_{xz} & Q_{xy}-Q_{yx} & Q_{xx}+Q_{yy}+Q_{zz} \end{bmatrix},$$

and find the eigenvector (x,y,z,w) corresponding to the largest eigenvalue (that value will be 1 if and only if Q is a pure rotation). The quaternion so obtained will correspond to the rotation closest to the original matrix Q[dubious – discuss][4]

Pairs of Unit Quaternions as Rotations in 4D Space

A pair of unit quaternions z_l and z_r can represent any rotation in 4D space. Given a four dimensional vector $\vec{v}$, and pretending that it is a quaternion, we can rotate the vector like this:

It is straightforward to check that for each matrix $M M^T = I$, that is, that each matrix (and hence both matrices together) represents a rotation. Note that since , the two matrices must commute. Therefore, there are two commuting subgroups of the set of four dimensional rotations. Arbitrary four dimensional rotations have 6 degrees of freedom, each matrix represents 3 of those 6 degrees of freedom.

Since an infinitesimal four-dimensional rotation can be represented by a pair of quaternions (as follows), all (non-infinitesimal) four-dimensional rotations can also be represented.

$$z_l\vec{v}z_r = \begin{pmatrix} 1 & -dt_{ab} & -dt_{ac} & -dt_{ad} \\ dt_{ab} & 1 & -dt_{bc} & -dt_{bd} \\ dt_{ac} & dt_{bc} & 1 & -dt_{cd} \\ dt_{ad} & dt_{bd} & dt_{cd} & 1 \end{pmatrix}\begin{pmatrix} w \\ x \\ y \\ z \end{pmatrix}$$

$$z_l = 1 + \frac{dt_{ab}+dt_{cd}}{2}i + \frac{dt_{ac}-dt_{bd}}{2}j + \frac{dt_{ad}+dt_{bc}}{2}k$$

$$z_r = 1 + \frac{dt_{ab}-dt_{cd}}{2}i + \frac{dt_{ac}+dt_{bd}}{2}j + \frac{dt_{ad}-dt_{bc}}{2}k$$

Conversion Between Quaternions and Euler Angles

Spatial rotations in three dimensions can be parametrized using both Euler angles and unit quaternions. This article explains how to convert between the two representations. Actually this simple use of "quaternions" was first presented by Euler some seventy years earlier than Hamilton to solve the problem of magic squares. For this reason the dynamics community commonly refers to quaternions in this application as "Euler parametres".

Definition

A unit quaternion can be described as:

$$\mathbf{q} = \begin{bmatrix} q_0 & q_1 & q_2 & q_3 \end{bmatrix}^T$$

$$|\mathbf{q}|^2 = q_0^2 + q_1^2 + q_2^2 + q_3^2 = 1$$

We can associate a quaternion with a rotation around an axis by the following expression

$$\mathbf{q}_0 = \cos(\alpha/2)$$

$$\mathbf{q}_1 = \sin(\alpha/2)\cos(\beta_x)$$

$$\mathbf{q}_2 = \sin(\alpha/2)\cos(\beta_y)$$

$$\mathbf{q}_3 = \sin(\alpha/2)\cos(\beta_z)$$

where α is a simple rotation angle (the value in radians of the angle of rotation) and $\cos(\beta_x)$, $\cos(\beta_y)$ and $\cos(\beta_z)$ are the "direction cosines" locating the axis of rotation (Euler's Theorem).

Rotation Matrices

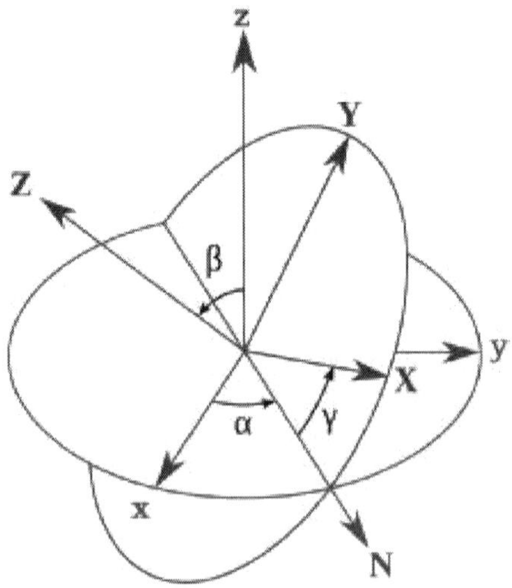

Figure: *Euler angles – The xyz (fixed) system is shown in blue, the XYZ (rotated) system is shown in red. The line of nodes, labelled N, is shown in green.*

The orthogonal matrix (post-multiplying a column vector) corresponding to a clockwise/left-handed rotation by the unit quaternion $q = q_0 + iq_1 + jq_2 + kq_3$ is given by the inhomogeneous expression

$$\begin{bmatrix} 1-2(q_2^2+q_3^2) & 2(q_1q_2-q_0q_3) & 2(q_0q_2+q_1q_3) \\ 2(q_1q_2+q_0q_3) & 1-2(q_1^2+q_3^2) & 2(q_2q_3-q_0q_1) \\ 2(q_1q_3-q_0q_2) & 2(q_0q_1+q_2q_3) & 1-2(q_1^2+q_2^2) \end{bmatrix}$$

or equivalently, by the homogeneous expression

$$\begin{bmatrix} q_0^2+q_1^2-q_2^2-q_3^2 & 2(q_1q_2-q_0q_3) & 2(q_0q_2+q_1q_3) \\ 2(q_1q_2+q_0q_3) & q_0^2-q_1^2+q_2^2-q_3^2 & 2(q_2q_3-q_0q_1) \\ 2(q_1q_3-q_0q_2) & 2(q_0q_1+q_2q_3) & q_0^2-q_1^2-q_2^2+q_3^2 \end{bmatrix}$$

If $q_0 + iq_1 + jq_2 + kq_3$ is not a unit quaternion then the homogeneous form is still a scalar multiple of a rotation matrix, while the inhomogeneous form is in general no longer an orthogonal matrix. This is why in numerical work the homogeneous form is to be preferred if distortion is to be avoided.

The orthogonal matrix (post-multiplying a column vector) corresponding to a clockwise/left-handed rotation with Euler angles φ, θ, ψ, with *x-y-z* convention, is given by:

$$\begin{bmatrix} \cos\theta\cos\psi & -\cos\phi\sin\psi+\sin\phi\sin\theta\cos\psi & \sin\phi\sin\psi+\cos\phi\sin\theta\cos\psi \\ \cos\theta\sin\psi & \cos\phi\cos\psi+\sin\phi\sin\theta\sin\psi & -\sin\phi\cos\psi+\cos\phi\sin\theta\sin\psi \\ -\sin\theta & \sin\phi\cos\theta & \cos\phi\cos\theta \end{bmatrix}$$

Conversion

By combining the quaternion representations of the Euler rotations we get

$$\mathbf{q} = R_z(\psi)R_y(\theta)R_x(\phi) = [\cos(\psi/2)+\mathbf{k}\sin(\psi/2)][\cos(\theta/2) \\ +\mathbf{j}\sin(\theta/2)][\cos(\phi/2)+\mathbf{i}\sin(\phi/2)]$$

$$\mathbf{q} = \begin{bmatrix} \cos(\phi/2)\cos(\theta/2)\cos(\psi/2)+\sin(\phi/2)\sin(\theta/2)\sin(\psi/2) \\ \sin(\phi/2)\cos(\theta/2)\cos(\psi/2)-\cos(\phi/2)\sin(\theta/2)\sin(\psi/2) \\ \cos(\phi/2)\sin(\theta/2)\cos(\psi/2)+\sin(\phi/2)\cos(\theta/2)\sin(\psi/2) \\ \cos(\phi/2)\cos(\theta/2)\sin(\psi/2)-\sin(\phi/2)\sin(\theta/2)\cos(\psi/2) \end{bmatrix}$$

For Euler angles we get:

$$\begin{bmatrix} \phi \\ \theta \\ \psi \end{bmatrix} = \begin{bmatrix} \arctan \dfrac{2(q_0q_1 + q_2q_3)}{1-2(q_1^2+q_2^2)} \\ \arcsin(2(q_0q_2 - q_3q_1)) \\ \arctan \dfrac{2(q_0q_3 + q_1q_2)}{1-2(q_2^2+q_3^2)} \end{bmatrix}$$

arctan and arcsin have a result between $-\pi/2$ and $\pi/2$. With three rotations between $-\pi/2$ and $\pi/2$ you can't have all possible orientations. You need to replace the arctan by atan2 to generate all the orientations.

$$\begin{bmatrix} \phi \\ \theta \\ \psi \end{bmatrix} = \begin{bmatrix} \text{atan2}(2(q_0q_1 + q_2q_3), 1-2(q_1^2+q_2^2)) \\ \arcsin(2(q_0q_2 - q_3q_1)) \\ \text{atan2}(2(q_0q_3 + q_1q_2), 1-2(q_2^2+q_3^2)) \end{bmatrix}$$

Singularities

One must be aware of singularities in the Euler angle parametrisation when the pitch approaches ±90° (north/south pole). These cases must be handled specially. The common name for this situation is gimbal lock.

Generalisations

The rotation group generalises quite naturally to n-dimensional Euclidean space, R^n. The group of all proper and improper rotations in n dimensions is called the orthogonal group, $O(n)$, and the subgroup of proper rotations is called the special orthogonal group, $SO(n)$. In special relativity, one works in a 4-dimensional vector space, known as Minkowski space rather than 3-dimensional Euclidean space. Unlike Euclidean space, Minkowski space has an inner product with an indefinite signature. However, one can still define *generalised rotations* which preserve this inner product. Such generalised rotations are known as Lorentz transformations and the group of all such transformations is called the Lorentz group. The rotation group SO(3) can be described as a subgroup of $E^+(3)$, the Euclidean group of direct isometries of R^3. This larger group is the group of all motions of a rigid body: each of these is a combination of a rotation about an arbitrary axis and a translation along the axis, or put differently, a combination of an element of SO(3) and an arbitrary translation.

In general, the rotation group of an object is the symmetry group within the group of direct isometries; in other words, the intersection of the full symmetry group and the group of direct isometries. For chiral objects it is the same as the full symmetry group.

Chapter

8

Rotation Matrix

In linear algebra, a rotation matrix is a matrix that is used to perform a rotation in Euclidean space. For example the matrix

$$R = \begin{bmatrix} \cos\theta & -\sin\theta \\ \sin\theta & \cos\theta \end{bmatrix}$$

rotates points in the xy-Cartesian plane counterclockwise through an angle θ about the origin of the Cartesian coordinate system. To perform the rotation using a rotation matrix R, the position of each point must be represented by a column vector v, containing the coordinates of the point. A rotated vector is obtained by using the matrix multiplication Rv. Since matrix multiplication has no effect on the zero vector (i.e., on the coordinates of the origin), rotation matrices can only be used to describe rotations about the origin of the coordinate system.

Rotation matrices provide a simple algebraic description of such rotations, and are used extensively for computations in geometry, physics, and computer graphics. In two-dimensional space, a rotation can be simply described by an angle θ of rotation, but it can be also represented by the four entries of a rotation matrix with two rows and two columns. In three-dimensional space, every rotation can be interpreted as a rotation by a given angle about a single fixed axis of rotation, and hence it can be simply described by an angle and a vector with three entries. However, it can also be represented by the nine entries of a rotation matrix with three rows and three columns. The notion of rotation is not commonly used in dimensions higher than 3; there is a notion of a rotational displacement, which can be represented by a matrix, but not associated single axis or angle.

Rotation matrices are square matrices, with real entries. More specifically they can be characterised as orthogonal matrices with determinant 1:

$$R^T = R^{-1}, \det R = 1$$

The set of all such matrices of size n forms a (generally not commutative) group, known as the special orthogonal group SO(n).

In some literature, the term *rotation* is generalised to include improper rotations, characterised by orthogonal matrices with determinant -1 (instead of +1). These combine proper rotations with *reflections* (which invert orientation). In this sense the members of the (general) orthogonal group O(n) may also be called rotation matrices. In other cases, where reflections are not being considered, the label *proper* may be dropped unless demands of clarity make it wise to specify it. This is the reason it is more frequent to associate the term *rotation matrix* with members of SO(n) (as in the rest of this article) while it may also be found referring to members of O(n).

In Two Dimensions

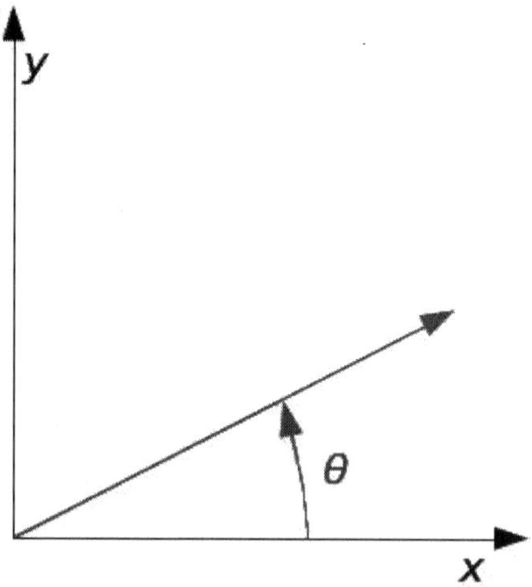

Figure: *A counterclockwise rotation of a vector through angle θ. The vector is initially aligned with the x-axis.*

In two dimensions every rotation matrix has the following form:

$$R(\theta) = \begin{bmatrix} \cos\theta & -\sin\theta \\ \sin\theta & \cos\theta \end{bmatrix}$$

This rotates column vectors by means of the following matrix multiplication:

$$\begin{bmatrix} x' \\ y' \end{bmatrix} = \begin{bmatrix} \cos\theta & -\sin\theta \\ \sin\theta & \cos\theta \end{bmatrix} \begin{bmatrix} x \\ y \end{bmatrix}$$

So the coordinates (x',y') of the point (x,y) after rotation are:

$$x' = x\cos\theta - y\sin\theta ,$$

$$y' = x\sin\theta + y\cos\theta .$$

The direction of vector rotation is counterclockwise if θ is positive (e.g. 90°), and clockwise if θ is negative (e.g. -90°).

$$R(-\theta) = \begin{bmatrix} \cos\theta & \sin\theta \\ -\sin\theta & \cos\theta \end{bmatrix}$$

Note that the two-dimensional case is the only non-trivial (e.g. one dimension) case where the rotation matrices group is commutative, so that it does not matter the order in which multiple rotations are performed.

Non-standard Orientation of the Coordinate System

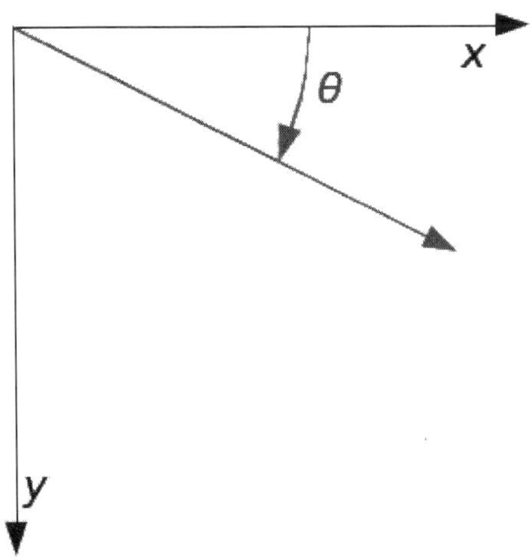

Figure: *A rotation through angle θ with non-standard axes.*

If a standard right-handed Cartesian coordinate system is used, with the x axis to the right and the y axis up, the rotation R(θ) is counterclockwise. If a left-handed Cartesian coordinate system is used, with x directed to the right but y directed down, R(θ) is clockwise. Such non-standard orientations are rarely used in mathematics but are common in 2D computer graphics, which often have the origin in the top left corner and the y-axis down the screen or page.

See below for other alternative conventions which may change the sense of the rotation produced by a rotation matrix.

Common Rotations

Particularly useful are the matrices for 90° and 180° rotations:

$$R(90^\circ) = \begin{bmatrix} 0 & -1 \\[3pt] 1 & 0 \end{bmatrix}$$ (90° counterclockwise rotation)

$$R(180^\circ) = \begin{bmatrix} -1 & 0 \\[3pt] 0 & -1 \end{bmatrix}$$ (180° rotation in either direction – a half-turn)

$$R(270^\circ) = \begin{bmatrix} 0 & 1 \\[3pt] -1 & 0 \end{bmatrix}$$ (270° counterclockwise rotation, the same as a 90° clockwise rotation)

Ambiguities

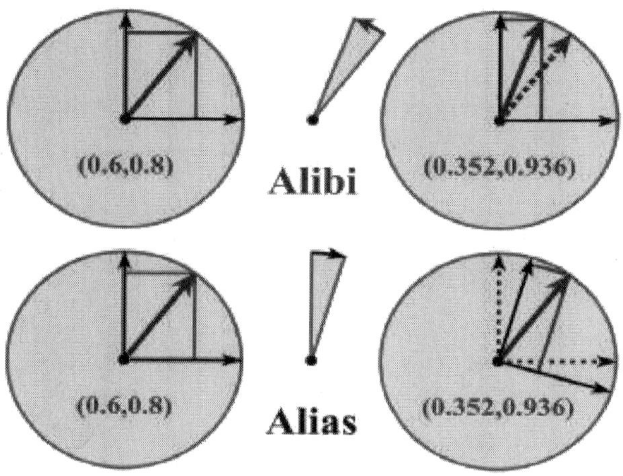

Figure: *Alias and alibi rotations*

The interpretation of a rotation matrix can be subject to many ambiguities.

Alias or Alibi Transformation: The change in a vector's coordinates can be due to a turn of the coordinate system (alias) or a turn of the vector (alibi). Any rotation can be legitimately described both ways, as vectors and coordinate systems actually rotate with respect to each other. Throughout this article, we chose the alibi approach to describe rotations.

Pre-multiplication or Post-multiplication: The vector can be pre-multiplied by a rotation matrix (Rv, where v is a column vector), or post-multiplied by it (vR, where v is a row vector). Throughout this article, we described rotations produced by means of a pre-multiplication.

Right- or Left-handed Coordinates: The matrix and the vector can be represented with respect to a right-handed or left-handed coordinate system. Throughout the article, we assumed a right-handed orientation, unless otherwise specified.

Vectors or Forms: The vector space has a dual space of linear forms, and the matrix can act on either vectors or forms.

In most cases the effect of the ambiguity is equivalent to the effect of a transposition of the rotation matrix.

Decompositions

Independent Planes

Consider the 3×3 rotation matrix

$$Q = \begin{bmatrix} 0.36 & 0.48 & -0.8 \\ -0.8 & 0.60 & 0 \\ 0.48 & 0.64 & 0.60 \end{bmatrix}.$$

If Q acts in a certain direction, v, purely as a scaling by a factor λ, then we have

$$Q\mathbf{v} = \lambda\mathbf{v},$$

so that

$$\mathbf{0} = (\lambda I - Q)\mathbf{v}.$$

Thus λ is a root of the characteristic polynomial for Q,

$$\begin{aligned} 0 &= \det(\lambda I - Q) \\ &= \lambda^3 - \tfrac{39}{25}\lambda^2 + \tfrac{39}{25}\lambda - 1 \\ &= (\lambda - 1)(\lambda^2 - \tfrac{14}{25}\lambda + 1). \end{aligned}$$

Two features are noteworthy. First, one of the roots (or eigenvalues) is 1, which tells us that some direction is unaffected by the matrix. For rotations in three dimensions, this is the *axis* of the rotation (a concept that has no meaning in any other dimension). Second, the other two roots are a pair of complex conjugates, whose product is 1 (the constant term of the quadratic), and whose sum is 2 cos θ (the negated linear term). This factorisation is of interest for 3 × 3 rotation matrices because the same thing occurs for all of them. (As special cases, for a null rotation the "complex conjugates" are both 1, and for a 180° rotation they are both "1.) Furthermore, a similar factorisation holds for any $n \times n$ rotation matrix. If the dimension, n, is odd, there will be a "dangling" eigenvalue of 1; and for any dimension the rest of the polynomial factors into quadratic terms like the one here (with the two special cases noted). We are guaranteed that the characteristic polynomial will have degree n and thus n eigenvalues. And since a rotation matrix commutes with its transpose, it is a normal matrix, so can be diagonalised. We conclude that every rotation matrix, when expressed in a suitable coordinate system, partitions into independent rotations of two-dimensional subspaces, at most $^{n}\mathrm{D}_2$ of them.

The sum of the entries on the main diagonal of a matrix is called the trace; it does not change if we reorient the coordinate system, and always equals the sum of the eigenvalues. This has the convenient implication for 2×2 and 3×3 rotation matrices that the trace reveals the angle of rotation, θ, in the two-dimensional (sub-)space. For a 2×2 matrix the trace is $2 \cos(\theta)$, and for a 3×3 matrix it is $1 + 2 \cos(\theta)$. In the three-dimensional case, the subspace consists of all vectors perpendicular to the rotation axis (the invariant direction, with eigenvalue 1). Thus we can extract from any 3×3 rotation matrix a rotation axis and an angle, and these completely determine the rotation.

Sequential Angles

The constraints on a 2×2 rotation matrix imply that it must have the form

$$Q = \begin{bmatrix} a & -b \\ b & a \end{bmatrix}$$

with $a^2+b^2 = 1$. Therefore we may set $a = \cos\theta$ and $b = \sin\theta$, for some angle θ. To solve for θ it is not enough to look at a alone or b alone; we must consider both together to place the angle in the correct quadrant, using a two-argument arctangent function.

Now consider the first column of a 3×3 rotation matrix,

$$\begin{bmatrix} a \\ b \\ c \end{bmatrix}.$$

Although a^2+b^2 will probably not equal 1, but some value $r^2 < 1$, we can use a slight variation of the previous computation to find a so-called Givens rotation that transforms the column to

$$\begin{bmatrix} r \\ 0 \\ c \end{bmatrix},$$

zeroing b. This acts on the subspace spanned by the x and y axes. We can then repeat the process for the xz subspace to zero c. Acting on the full matrix, these two rotations produce the schematic form

$$Q_{xz}Q_{xy}Q = \begin{bmatrix} 1 & 0 & 0 \\ 0 & * & * \\ 0 & * & * \end{bmatrix}.$$

Shifting attention to the second column, a Givens rotation of the yz subspace can now zero the z value. This brings the full matrix to the form

$$Q_{yz}Q_{xz}Q_{xy}Q = \begin{bmatrix} 1 & 0 & 0 \\ 0 & 1 & 0 \\ 0 & 0 & 1 \end{bmatrix},$$

which is an identity matrix. Thus we have decomposed Q as

$$Q = Q_{xy}^{-1}Q_{xz}^{-1}Q_{yz}^{-1}.$$

An $n \times n$ rotation matrix will have $(n-1)+(n-2)+ \ldots -+2+1$, or

$$\sum_{k=1}^{n-1} k = \frac{n(n-1)}{2}$$

entries below the diagonal to zero. We can zero them by extending the same idea of stepping through the columns with a series of rotations in a fixed sequence of planes. We conclude that the set of $n \times n$ rotation matrices, each of which has n^2 entries, can be parameterised by $n(n-1)/2$ angles.

xzx_w	xzy_w	xyx_w	xyz_w
yxy_w	yxz_w	yzy_w	yzx_w
zyz_w	zyx_w	zxz_w	zxy_w
xzx_b	yzx_b	xyx_b	zyx_b
yxy_b	zxy_b	yzy_b	xzy_b
zyz_b	xyz_b	zxz_b	yxz_b

In three dimensions this restates in matrix form an observation made by Euler, so mathematicians call the ordered sequence of three angles Euler angles. However, the situation is somewhat more complicated than we have so far indicated. Despite the small dimension, we actually have considerable freedom in the sequence of axis pairs we use; and we also have some freedom in the choice of angles.

Thus we find many different conventions employed when three-dimensional rotations are parameterized for physics, or medicine, or chemistry, or other disciplines.

When we include the option of world axes or body axes, 24 different sequences are possible. And while some disciplines call any sequence Euler angles, others give different names (Euler, Cardano, Tait-Bryan, Roll-pitch-yaw) to different sequences. One reason for the large number of options is that, as noted previously, rotations in three dimensions (and higher) do not commute.

If we reverse a given sequence of rotations, we get a different outcome. This also implies that we cannot compose two rotations by adding their corresponding angles. Thus *Euler angles are not vectors*, despite a similarity in appearance as a triple of numbers.

Nested Dimensions]

A 3×3 rotation matrix like

$$Q_{3\times 3} = \begin{bmatrix} \cos\theta & -\sin\theta & 0 \\ \sin\theta & \cos\theta & 0 \\ 0 & 0 & 1 \end{bmatrix}$$

suggests a 2×2 rotation matrix,

$$Q_{2\times 2} = \begin{bmatrix} \cos\theta & -\sin\theta \\ \sin\theta & \cos\theta \end{bmatrix},$$

is embedded in the upper left corner:

$$Q_{3\times 3} = \begin{bmatrix} Q_{2\times 2} & \mathbf{0} \\ \mathbf{0}^T & 1 \end{bmatrix}.$$

This is no illusion; not just one, but many, copies of n-dimensional rotations are found within $(n+1)$-dimensional rotations, as subgroups. Each embedding leaves one direction fixed, which in the case of 3×3 matrices is the rotation axis. For example, we have

$$Q_x(\theta) = \begin{bmatrix} 1 & 0 & 0 \\ 0 & \cos\theta & -\sin\theta \\ 0 & \sin\theta & \cos\theta \end{bmatrix},$$

$$Q_y(\theta) = \begin{bmatrix} \cos\theta & 0 & \sin\theta \\ 0 & 1 & 0 \\ -\sin\theta & 0 & \cos\theta \end{bmatrix},$$

$$Q_z(\theta) = \begin{bmatrix} \cos\theta & -\sin\theta & 0 \\ \sin\theta & \cos\theta & 0 \\ 0 & 0 & 1 \end{bmatrix},$$

fixing the x axis, the y axis, and the z axis, respectively. The rotation axis need not be a coordinate axis; if u = (x,y,z) is a unit vector in the desired direction, then

$$\begin{aligned} Q_{\mathbf{u}}(\theta) &= \begin{bmatrix} 0 & -z & y \\ z & 0 & -x \\ -y & x & 0 \end{bmatrix} \sin\theta + (I - \mathbf{u}\mathbf{u}^T)\cos\theta + \mathbf{u}\mathbf{u}^T \\ &= \begin{bmatrix} (1-x^2)c_\theta + x^2 & -zs_\theta - xyc_\theta + xy & ys_\theta - xzc_\theta + xz \\ zs_\theta - xyc_\theta + xy & (1-y^2)c_\theta + y^2 & -xs_\theta - yzc_\theta + yz \\ -ys_\theta - xzc_\theta + xz & xs_\theta - yzc_\theta + yz & (1-z^2)c_\theta + z^2 \end{bmatrix} \\ &= \begin{bmatrix} x^2(1-c_\theta) + c_\theta & xy(1-c_\theta) - zs_\theta & xz(1-c_\theta) + ys_\theta \\ xy(1-c_\theta) + zs_\theta & y^2(1-c_\theta) + c_\theta & yz(1-c_\theta) - xs_\theta \\ xz(1-c_\theta) - ys_\theta & yz(1-c_\theta) + xs_\theta & z^2(1-c_\theta) + c_\theta \end{bmatrix}, \end{aligned}$$

where $c_\theta = \cos\theta$, $s_\theta = \sin\theta$, is a rotation by angle θ leaving axis u fixed.

A direction in $(n+1)$-dimensional space will be a unit magnitude vector, which we may consider a point on a generalised sphere, S^n. Thus it is natural to describe the rotation group $SO(n+1)$ as combining $SO(n)$ and S^n. A suitable formalism is the fibre bundle,

$$SO(n) \to SO(n+1) \to S^n,$$

where for every direction in the "base space", S^n, the "fibre" over it in the "total space", $SO(n+1)$, is a copy of the "fibre space", $SO(n)$, namely the rotations that keep that direction fixed.

Thus we can build an $n \times n$ rotation matrix by starting with a 2×2 matrix, aiming its fixed axis on S^2 (the ordinary sphere in three-dimensional space), aiming the resulting rotation on S^3, and so on up through S^{n-1}. A point on S^n can be selected using n numbers, so we again have $n(n-1)/2$ numbers to describe any $n \times n$ rotation matrix.

In fact, we can view the sequential angle decomposition, discussed previously, as reversing this process. The composition of $n-1$ Givens rotations brings the first column (and row) to $(1,0,\ldots,0)$, so that the remainder of the matrix is a rotation matrix of dimension one less, embedded so as to leave (1,0,…,0) fixed.

Skew Parametres via Cayley's Formula

When an $n\times n$ rotation matrix, Q, does not include –1 as an eigenvalue, so that none of the planar rotations of which it is composed are 180° rotations, then $Q+I$ is an invertible matrix. Most rotation matrices fit this description, and for them we can show that $(Q-I)(Q+I)^{-1}$ is a skew-symmetric matrix, A. Thus $A^T = -A$; and since the diagonal is necessarily zero, and since the upper triangle determines the lower one, A contains $n(n-1)/2$ independent numbers. Conveniently, $I-A$ is invertible whenever A is skew-symmetric; thus we can recover the original matrix using the *Cayley transform*,

$$A \mapsto (I+A)(I-A)^{-1},$$

which maps any skew-symmetric matrix A to a rotation matrix. In fact, aside from the noted exceptions, we can produce any rotation matrix in this way. Although in practical applications we can hardly afford to ignore 180° rotations, the Cayley transform is still a potentially useful tool, giving a parameterisation of most rotation matrices without trigonometric functions.

In three dimensions, for example, we have (Cayley 1846)

$$\begin{bmatrix} 0 & -z & y \\ z & 0 & -x \\ -y & x & 0 \end{bmatrix} \mapsto$$

$$\frac{1}{1+x^2+y^2+z^2}\begin{bmatrix} 1+x^2-y^2-z^2 & 2xy-2z & 2y+2xz \\ 2xy+2z & 1-x^2+y^2-z^2 & 2yz-2x \\ 2xz-2y & 2x+2yz & 1-x^2-y^2+z^2 \end{bmatrix}.$$

If we condense the skew entries into a vector, (x,y,z), then we produce a 90° rotation around the x axis for (1,0,0), around the y axis for (0,1,0), and around the z axis for (0,0,1). The 180° rotations are just out of reach; for, in the limit as x goes to infinity, $(x,0,0)$ does approach a 180° rotation around the x axis, and similarly for other directions.

Lie Theory

Lie Group

We have established that $n\times n$ rotation matrices form a group, the special orthogonal group, SO(n). This algebraic structure is coupled with a topological structure, in that the operations of multiplication and taking the inverse (which here is merely transposition) are continuous functions of the matrix entries. Thus $SO(n)$ is a classic example of a topological group. (In purely topological terms, it is a compact manifold.) Furthermore, the operations are not only continuous, but smooth, so $SO(n)$ is a differentiable manifold and a Lie group (Baker (2003); Fulton & Harris (1991)).

Most properties of rotation matrices depend very little on the dimension, n; yet in Lie group theory we see systematic differences between even dimensions and odd dimensions. As well, there are some irregularities below $n = 5$; for example, $SO(4)$ is, anomalously, not a simple Lie group, but instead isomorphic to the product of S^3 and $SO(3)$.

Lie Algebra

Associated with every Lie group is a Lie algebra, a linear space equipped with a bilinear alternating product called a bracket. The algebra for $SO(n)$ is denoted by

$$SO(n),$$

and consists of all skew-symmetric $n \times n$ matrices (as implied by differentiating the orthogonality condition, $I = Q^T Q$). The bracket, $[A_1, A_2]$, of two skew-symmetric matrices is defined to be $A_1A_2 - A_2A_1$, which is again a skew-symmetric matrix. This Lie algebra bracket captures the essence of the Lie group product via infinitesimals.

For 2×2 rotation matrices, the Lie algebra is a one-dimensional vector space, multiples of

$$J = \begin{bmatrix} 0 & -1 \\ 1 & 0 \end{bmatrix}.$$

Here the bracket always vanishes, which tells us that, in two dimensions, rotations commute. Not so in any higher dimension. For 3×3 rotation matrices, we have a three-dimensional vector space with the convenient basis

$$A_x = \begin{bmatrix} 0 & 0 & 0 \\ 0 & 0 & -1 \\ 0 & 1 & 0 \end{bmatrix}, \quad A_y = \begin{bmatrix} 0 & 0 & 1 \\ 0 & 0 & 0 \\ -1 & 0 & 0 \end{bmatrix}, \quad A_z = \begin{bmatrix} 0 & -1 & 0 \\ 1 & 0 & 0 \\ 0 & 0 & 0 \end{bmatrix}.$$

The Lie brackets of these generators are as follows

$$[A_x, A_y] = A_z, \quad [A_z, A_x] = A_y, \quad [A_y, A_z] = A_x.$$

We can conveniently identify any matrix in this Lie algebra with a vector in R^3,

$$\begin{aligned} \omega &= (x, y, z) \\ \tilde{\omega} &= xA_x + yA_y + zA_z \\ &= \begin{bmatrix} 0 & -z & y \\ z & 0 & -x \\ -y & x & 0 \end{bmatrix}. \end{aligned}$$

Under this identification, the so(3) bracket has a memorable description; it is the vector cross product,

$$[\tilde{\mathbf{u}}, \tilde{\mathbf{v}}] = (\mathbf{u} \times \mathbf{v})^{\sim}.$$

The matrix identified with a vector v is also memorable, because

$$\tilde{\mathbf{v}}\mathbf{u} = \mathbf{v} \times \mathbf{u}.$$

Notice this implies that v is in the null space of the skew-symmetric matrix with which it is identified, because v×v is always the zero vector.

Exponential Map

Connecting the Lie algebra to the Lie group is the *exponential map*, which we define using the familiar power series for e^x (Wedderburn 1934, §8.02),

$$\begin{aligned} \exp : \mathfrak{so}(n) &\to SO(n) \\ A &\mapsto I + A + \tfrac{1}{2}A^2 + \tfrac{1}{6}A^3 + \cdots + \tfrac{1}{k!}A^k + \cdots \\ &= \sum_{k=0}^{\infty} \frac{1}{k!} A^k. \end{aligned}$$

For any skew-symmetric A, exp(A) is always a rotation matrix.

An important practical example is the 3×3 case, where we have seen we can identify every skew-symmetric matrix with a vector $\omega = \mathrm{u}\theta$, where u = (x,y,z) is a unit magnitude vector.

Recall that u is in the null space of the matrix associated with ω, so that if we use a basis with u as the z axis the final column and row will be zero. Thus we know in advance that the exponential matrix must leave u fixed. It is mathematically impossible to supply a straightforward formula for such a basis as a function of u (its existence would violate the hairy ball theorem), but direct exponentiation is possible, and yields

$$\exp(\tilde{\omega}) = \exp\left(\begin{bmatrix} 0 & -z\theta & y\theta \\ z\theta & 0 & -x\theta \\ -y\theta & x\theta & 0 \end{bmatrix}\right)$$

$$= \begin{bmatrix} 2(x^2-1)s^2+1 & 2xys^2-2zcs & 2xzs^2+2ycs \\ 2xys^2+2zcs & 2(y^2-1)s^2+1 & 2yzs^2-2xcs \\ 2xzs^2-2ycs & 2yzs^2+2xcs & 2(z^2-1)s^2+1 \end{bmatrix}$$

where $c = \cos \theta/2$, $s = \sin \theta/2$. We recognize this as our matrix for a rotation around axis u by angle θ. We also note that this mapping of skew-symmetric matrices is quite different from the Cayley transform discussed earlier.

In any dimension, if we choose some nonzero A and consider all its scalar multiples, exponentiation yields rotation matrices along a *geodesic* of the group manifold, forming a one-parametre subgroup of the Lie group. More broadly, the exponential map provides a homeomorphism between a neighbourhood of the origin in the Lie algebra and a neighbourhood of the identity in the Lie group. In fact, we can produce any rotation matrix as the exponential of some skew-symmetric matrix, so for these groups the exponential map is a *surjection.*

Baker–Campbell–Hausdorff Formula

Suppose we are given A and B in the Lie algebra. Their exponentials, $\exp(A)$ and $\exp(B)$, are rotation matrices, which we can multiply. Since the exponential map is a surjection, we know that for some C in the Lie algebra, $\exp(A)\exp(B) = \exp(C)$, and we write

$$A * B = C.$$

When $\exp(A)$ and $\exp(B)$ commute (which always happens for 2×2 matrices, but not higher), then $C = A+B$, mimicking the behaviour of complex exponentiation. The general case is given by the BCH formula, a series expanded in terms of the bracket (Hall 2004, Ch. 3; Varadarajan 1984, §2.15). For matrices, the bracket is the same operation as the commutator, which detects lack of commutativity in multiplication. The general formula begins as follows.

$$A * B = A + B + \tfrac{1}{2}[A,B] + \tfrac{1}{12}[A,[A,B]] - \tfrac{1}{12}[B,[A,B]] - \cdots$$

Representation of a rotation matrix as a sequential angle decomposition, as in Euler angles, may tempt us to treat rotations as a vector space, but the higher order terms in the BCH formula reveal that to be a mistake.

We again take special interest in the 3×3 case, where $[A,B]$ equals the cross product, $A\times B$. If A and B are linearly independent, then A, B, and $A\times B$ can be used as a basis; if not, then A and B commute. And conveniently, in this dimension the summation in the BCH formula has a closed form (Engø 2001) as $\alpha A+\beta B+\gamma(A\times B)$.

Spin Group

The Lie group of $n \times n$ rotation matrices, $SO(n)$, is a compact and path-connected manifold, and thus locally compact and connected. However, it is not simply connected, so Lie theory tells us it is a kind of "shadow" (a homomorphic image) of a universal covering group. Often the covering group, which in this case is the spin group denoted by Spin(n), is simpler and more natural to work with (Baker 2003, Ch. 5; Fulton & Harris 1991, pp. 299–315). In the case of planar rotations, $SO(2)$ is topologically a circle, S^1. Its universal covering group, Spin(2), is isomorphic to the real line, R, under addition. In other words, whenever we use angles of arbitrary magnitude, which we often do, we are essentially taking advantage of the convenience of the "mother space". Every 2×2 rotation matrix is produced by a countable infinity of angles, separated by integer multiples of 2π. Correspondingly, the fundamental group of $SO(2)$ is isomorphic to the integers, Z.

In the case of spatial rotations, $SO(3)$ is topologically equivalent to three-dimensional real projective space, RP^3. Its universal covering group, Spin(3), is isomorphic to the 3-sphere, S^3. Every 3×3 rotation matrix is produced by two opposite points on the sphere. Correspondingly, the fundamental group of $SO(3)$ is isomorphic to the two-element group, Z_2. We can also describe Spin(3) as isomorphic to quaternions of unit norm under multiplication, or to certain 4×4 real matrices, or to 2×2 complex special unitary matrices.

Concretely, a unit quaternion, q, with

$$\begin{aligned} q &= w + \mathbf{i}x + \mathbf{j}y + \mathbf{k}z, \\ 1 &= w^2 + x^2 + y^2 + z^2, \end{aligned}$$

produces the rotation matrix

$$Q = \begin{bmatrix} 1-2y^2-2z^2 & 2xy-2zw & 2xz+2yw \\ 2xy+2zw & 1-2x^2-2z^2 & 2yz-2xw \\ 2xz-2yw & 2yz+2xw & 1-2x^2-2y^2 \end{bmatrix}.$$

This is our third version of this matrix, here as a rotation around non-unit axis vector (x,y,z) by angle 2θ, where $\cos\theta = w$ and $|\sin\theta| = \|(x,y,z)\|$. (The proper sign for $\sin\theta$ is implied once the signs of the axis components are decided.) Many features of this case are the same for higher dimensions. The coverings are all two-to-one, with $SO(n)$, $n > 2$, having fundamental group Z_2. The natural setting for these groups is within a Clifford algebra. And the action of the rotations is produced by a kind of "sandwich", denoted by qvq^{*}.

Infinitesimal Rotations

The matrices in the Lie algebra are not themselves rotations; the skew-symmetric matrices are derivatives, proportional differences of rotations. An actual "differential rotation", or *infinitesimal rotation matrix* has the form

$$I + A d\theta,$$

where $d\theta$ is vanishingly small. These matrices do not satisfy all the same properties as ordinary finite rotation matrices under the usual treatment of infinitesimals (Goldstein, Poole & Safko 2002, §4.8). To understand what this means, consider

$$dA_x = \begin{bmatrix} 1 & 0 & 0 \\ 0 & 1 & -d\theta \\ 0 & d\theta & 1 \end{bmatrix}.$$

We first test the orthogonality condition, $Q^T Q = I$. The product is

$$dA_x^T dA_x = \begin{bmatrix} 1 & 0 & 0 \\ 0 & 1+d\theta^2 & 0 \\ 0 & 0 & 1+d\theta^2 \end{bmatrix},$$

differing from an identity matrix by second order infinitesimals, which we discard. So to first order, an infinitesimal rotation matrix is an orthogonal matrix. Next we examine the square of the matrix.

$$dA_x^2 = \begin{bmatrix} 1 & 0 & 0 \\ 0 & 1-d\theta^2 & -2d\theta \\ 0 & 2d\theta & 1-d\theta^2 \end{bmatrix}.$$

Again discarding second order effects, we see that the angle simply doubles. This hints at the most essential difference in behaviour, which we can exhibit with the assistance of a second infinitesimal rotation,

$$dA_y = \begin{bmatrix} 1 & 0 & d\phi \\ 0 & 1 & 0 \\ -d\phi & 0 & 1 \end{bmatrix}.$$

Compare the products $dA_x dA_y$ and $dA_y dA_x$.

$$dA_x dA_y = \begin{bmatrix} 1 & 0 & d\phi \\ d\theta d\phi & 1 & -d\theta \\ -d\phi & d\theta & 1 \end{bmatrix}$$

$$dA_y dA_x = \begin{bmatrix} 1 & d\theta d\phi & d\phi \\ 0 & 1 & -d\theta \\ -d\phi & d\theta & 1 \end{bmatrix}$$

Since $d\theta\, d\varphi$ is second order, we discard it; thus, to first order, multiplication of infinitesimal rotation matrices is commutative. In fact,

$$dA_x dA_y = dA_y dA_x,$$

again to first order. Put in other words, the order in which infinitesimal rotations are applied is irrelevant, this useful fact makes, for example, derivation of rigid body rotation relatively simple.

But we must always be careful to distinguish (the first order treatment of) these infinitesimal rotation matrices from both finite rotation matrices and from derivatives of rotation matrices (namely skew-symmetric matrices).

Contrast the behaviour of finite rotation matrices in the BCH formula with that of infinitesimal rotation matrices, where all the commutator terms will be second order infinitesimals so we *do* have a vector space.

Conversions

We have seen the existence of several decompositions that apply in any dimension, namely independent planes, sequential angles, and nested dimensions. In all these cases we can either decompose a matrix or construct one. We have also given special attention to 3×3 rotation matrices, and these warrant further attention, in both directions (Stuelpnagel 1964).

Quaternion

Given the unit quaternion $q = (w,x,y,z)$, the equivalent 3×3 rotation matrix is

$$Q = \begin{bmatrix} 1-2y^2-2z^2 & 2xy-2zw & 2xz+2yw \\ 2xy+2zw & 1-2x^2-2z^2 & 2yz-2xw \\ 2xz-2yw & 2yz+2xw & 1-2x^2-2y^2 \end{bmatrix}.$$

Now every quaternion component appears multiplied by two in a term of degree two, and if all such terms are zero what's left is an identity matrix. This leads to an efficient, robust conversion from any quaternion — whether unit, nonunit, or even zero — to a 3×3 rotation matrix.

$Nq = w\hat{}2 + x\hat{}2 + y\hat{}2 + z\hat{}2$
$X = x{*}s;\ Y = y{*}s;\ Z = z{*}s$
$wX = w{*}X;\ wY = w{*}Y;\ wZ = w{*}Z$
$xX = x{*}X;\ xY = x{*}Y;\ xZ = x{*}Z$
$[1.0\text{-}(yY{+}zZ)\ \ xY\text{-}wZ\ \ xZ{+}wY]$
$[xY{+}wZ\ \ 1.0\text{-}(xX{+}zZ)\ \ yZ\text{-}wX]$
$[xZ\text{-}wY\ \ yZ{+}wX\ \ 1.0\text{-}(xX{+}yY)]$

Freed from the demand for a unit quaternion, we find that nonzero quaternions act as homogeneous coordinates for 3×3 rotation matrices. The Cayley transform, discussed earlier, is obtained by scaling the quaternion so that its w component is 1. For a 180° rotation around any axis, w will be zero, which explains the Cayley limitation. The sum of the entries along the main diagonal (the trace), plus one, equals $4-4(x^2+y^2+z^2)$, which is $4w^2$. Thus we can write the trace itself as $2w^2+2w^2-1$; and from the previous version of the matrix we see that the diagonal entries themselves have the same form: $2x^2+2w^2-1$, $2y^2+2w^2-1$, and $2z^2+2w^2-1$. So we can easily compare the magnitudes of all four quaternion components using the matrix diagonal. We can, in fact, *obtain* all four magnitudes using sums and square roots, and choose consistent signs using the skew-symmetric part of the off-diagonal entries.

t = Q_{xx}+Q_{yy}+Q_{zz} (trace of Q)
r = sqrt(1+t)
w = 0.5*r
x = copysign(0.5*sqrt(1+Q_{xx}-Q_{yy}-Q_{zz}), Q_{zy}-Q_{yz})
y = copysign(0.5*sqrt(1-Q_{xx}+Q_{yy}-Q_{zz}), Q_{xz}-Q_{zx})
z = copysign(0.5*sqrt(1-Q_{xx}-Q_{yy}+Q_{zz}), Q_{yx}-Q_{xy})

where copysign(x,y) is x with the sign of y:

$$\text{copysign}(x, y) = \text{sign}(y)\ |x|.$$

Alternatively, use a single square root and division

t = Q_{xx}+Q_{yy}+Q_{zz}
r = sqrt(1+t)
s = 0.5/r
w = 0.5*r
x = (Q_{zy}-Q_{yz})*s
y = (Q_{xz}-Q_{zx})*s
z = (Q_{yx}-Q_{xy})*s

This is numerically stable so long as the trace, t, is not negative; otherwise, we risk dividing by (nearly) zero. In that case, suppose Q_{xx} is the largest diagonal entry, so x will have the largest magnitude (the other cases are similar); then the following is safe.

$t = Q_{xx}+Q_{yy}+Q_{zz}$
$r = \text{sqrt}(1+Q_{xx}-Q_{yy}-Q_{zz})$
$s = 0.5/r$
$w = (Q_{zy}-Q_{yz})*s$
$x = 0.5*r$
$y = (Q_{xy}+Q_{yx})*s$
$z = (Q_{zx}+Q_{xz})*s$

If the matrix contains significant error, such as accumulated numerical error, we may construct a symmetric 4×4 matrix,

$$K = \frac{1}{3}\begin{bmatrix} Q_{xx}-Q_{yy}-Q_{zz} & Q_{yx}+Q_{xy} & Q_{zx}+Q_{xz} & Q_{yz}-Q_{zy} \\ Q_{yx}+Q_{xy} & Q_{yy}-Q_{xx}-Q_{zz} & Q_{zy}+Q_{yz} & Q_{zx}-Q_{xz} \\ Q_{zx}+Q_{xz} & Q_{zy}+Q_{yz} & Q_{zz}-Q_{xx}-Q_{yy} & Q_{xy}-Q_{yx} \\ Q_{yz}-Q_{zy} & Q_{zx}-Q_{xz} & Q_{xy}-Q_{yx} & Q_{xx}+Q_{yy}+Q_{zz} \end{bmatrix},$$

and find the eigenvector, (w,x,y,z), of its largest magnitude eigenvalue. (If Q is truly a rotation matrix, that value will be 1.) The quaternion so obtained will correspond to the rotation matrix closest to the given matrix (Bar-Itzhack 2000).

Polar Decomposition

If the $n{\times}n$ matrix M is non-singular, its columns are linearly independent vectors; thus the Gram–Schmidt process can adjust them to be an orthonormal basis. Stated in terms of numerical linear algebra, we convert M to an orthogonal matrix, Q, using QR decomposition. However, we often prefer a Q "closest" to M, which this method does not accomplish. For that, the tool we want is the polar decomposition (Fan & Hoffman 1955; Higham 1989). To measure closeness, we may use any matrix norm invariant under orthogonal transformations. A convenient choice is the Frobenius norm, $\|Q-M\|_F$, squared, which is the sum of the squares of the element differences. Writing this in terms of the trace, Tr, our goal is,

Find Q minimising $\text{Tr}((Q-M)^T(Q-M))$, subject to $Q^TQ = I$.

Though written in matrix terms, the objective function is just a quadratic polynomial. We can minimise it in the usual way, by finding where its derivative is zero. For a 3×3 matrix, the orthogonality constraint implies six scalar equalities that the entries of Q must satisfy. To incorporate the constraint(s), we may employ a standard technique, Lagrange multipliers, assembled as a symmetric matrix, Y. Thus our method is:

Differentiate Tr($(Q-M)^T(Q-M) + (Q^TQ-I)Y$) with respect to (the entries of) Q, and equate to zero.

Consider a 2×2 example. Including constraints, we seek to minimise

$$
\begin{aligned}
&(Q_{xx} - M_{xx})^2 + (Q_{xy} - M_{xy})^2 \\
&+(Q_{yx} - M_{yx})^2 + (Q_{yy} - M_{yy})^2 \\
&+(Q_{xx}^2 + Q_{yx}^2 - 1)Y_{xx} + (Q_{xy}^2 + Q_{yy}^2 - 1)Y_{yy} \\
&+2(Q_{xx}Q_{xy} + Q_{yx}Q_{yy})Y_{xy}.
\end{aligned}
$$

Taking the derivative with respect to Q_{xx}, Q_{xy}, Q_{yx}, Q_{yy} in turn, we assemble a matrix.

$$
2\begin{bmatrix} Q_{xx} - M_{xx} + Q_{xx}Y_{xx} + Q_{xy}Y_{xy} & Q_{xy} - M_{xy} + Q_{xx}Y_{xy} + Q_{xy}Y_{yy} \\ Q_{yx} - M_{yx} + Q_{yx}Y_{xx} + Q_{yy}Y_{xy} & Q_{yy} - M_{yy} + Q_{yx}Y_{xy} + Q_{yy}Y_{yy} \end{bmatrix}
$$

In general, we obtain the equation

$$0 = 2(Q - M) + 2QY,$$

so that

$$M = Q(I + Y) = QS,$$

where Q is orthogonal and S is symmetric. To ensure a minimum, the Y matrix (and hence S) must be positive definite. Linear algebra calls QS the polar decomposition of M, with S the positive square root of $S^2 = M^TM$.

$$S^2 = (Q^TM)^T(Q^TM) = M^TQQ^TM = M^TM$$

When M is non-singular, the Q and S factors of the polar decomposition are uniquely determined. However, the determinant of S is positive because S is positive definite, so Q inherits the sign of the determinant of M. That is, Q is only guaranteed to be orthogonal, not a rotation matrix. This is unavoidable; an M with negative determinant has no uniquely defined closest rotation matrix.

Axis and Angle

To efficiently construct a rotation matrix Q from an angle θ and a unit axis u, we can take advantage of symmetry and skew-symmetry within the entries. If x, y, and z are the components of the unit vector representing the axis, and

$$c = \cos(\theta);\ s = \sin(\theta);\ C = 1\text{-}c$$

then Q is

[$xxC+c$ $xyC-zs$ $xzC+ys$]

[$yxC+zs$ $yyC+c$ $yzC-xs$]

[$zxC-ys$ $zyC+xs$ $zzC+c$]

Determining an axis and angle, like determining a quaternion, is only possible up to sign; that is, (u,θ) and (“u,–θ) correspond to the same rotation matrix, just like q and “q. As well, axis-angle extraction presents additional difficulties. The angle can be restricted to be from 0° to 180°, but angles are formally ambiguous by multiples of 360°. When the angle is zero, the axis is undefined. When the angle is 180°, the matrix becomes symmetric, which has implications in extracting the axis. Near multiples of 180°, care is needed to avoid numerical problems: in extracting the angle, a two-argument arctangent with atan2(sin θ,cos θ) equal to θ avoids the insensitivity of arccosine; and in computing the axis magnitude to force unit magnitude, a brute-force approach can lose accuracy through underflow (Moler & Morrison 1983).

A partial approach is as follows.

$x = Q_{zy}-Q_{yz}$

$y = Q_{xz}-Q_{zx}$

$z = Q_{yx}-Q_{xy}$

$r = \mathrm{sqrt}(x^2 + y^2 + z^2)$

$t = Q_{xx}+Q_{yy}+Q_{zz}$ (trace of matrix Q)

$\theta = \mathrm{atan2}(r,t-1)$

The x, y, and z components of the axis would then be divided by r. A fully robust approach will use different code when the trace t is negative, as with quaternion extraction. When r is zero because the angle is zero, an axis must be provided from some source other than the matrix.

Euler Angles

Complexity of conversion escalates with Euler angles (used here in the broad sense). The first difficulty is to establish which of the twenty-four variations of Cartesian axis order we will use. Suppose the three angles are θ_1, θ_2, θ_3; physics and chemistry may interpret these as

$$Q(\theta_1,\theta_2,\theta_3) = Q_x(\theta_1)Q_y(\theta_2)Q_z(\theta_3),$$

while aircraft dynamics may use

$$Q(\theta_1,\theta_2,\theta_3) = Q_z(\theta_3)Q_y(\theta_2)Q_x(\theta_1).$$

One systematic approach begins with choosing the right-most axis. Among all permutations of (x,y,z), only two place that axis first; one is an even permutation and the other odd. Choosing parity thus establishes the middle axis. That leaves two choices for the left-most axis, either duplicating the first or not. These three choices gives us 3×2×2 = 12 variations; we double that to 24 by choosing static or rotating axes.

This is enough to construct a matrix from angles, but triples differing in many ways can give the same rotation matrix. For example, suppose we use the zyz convention above; then we have the following equivalent pairs:

(90°, 45°, – 105°) a – (– 270°, – 315°, 255°) *multiples of 360°*

(72°, 0°, 0°) a – (40°, 0°, 32°) *singular alignment*

(45°, 60°, – 30°) a – (– 135°, – 60°, 150°) *bistable flip*

Angles for any order can be found using a concise common routine (Herter & Lott 1993; Shoemake 1994).

The problem of singular alignment, the mathematical analog of physical gimbal lock, occurs when the middle rotation aligns the axes of the first and last rotations. It afflicts every axis order at either even or odd multiples of 90°. These singularities are not characteristic of the rotation matrix as such, and only occur with the usage of Euler angles.

The singularities are avoided when considering and manipulating the rotation matrix as orthonormal row vectors (in 3D applications often named 'right'-vector, 'up'-vector and 'out'-vector) instead of as angles. The singularities are also avoided when working with quaternions.

Chapter

9

Uniform Random Rotation Matrices

We sometimes need to generate a uniformly distributed random rotation matrix. It seems intuitively clear in two dimensions that this means the rotation angle is uniformly distributed between 0 and 2π. That intuition is correct, but does not carry over to higher dimensions. For example, if we decompose 3×3 rotation matrices in axis-angle form, the angle should *not* be uniformly distributed; the probability that (the magnitude of) the angle is at most θ should be ${}^{1}D_{\pi}(\theta - \sin\theta)$, for $0 \leq \theta \leq \pi$.

Since SO(n) is a connected and locally compact Lie group, we have a simple standard criterion for uniformity, namely that the distribution be unchanged when composed with any arbitrary rotation (a Lie group "translation"). This definition corresponds to what is called *Haar measure*. León, Massé & Rivest (2006) show how to use the Cayley transform to generate and test matrices according to this criterion.

We can also generate a uniform distribution in any dimension using the *subgroup algorithm* of Diaconis & Shashahani (1987). This recursively exploits the nested dimensions group structure of SO(n), as follows. Generate a uniform angle and construct a 2×2 rotation matrix. To step from n to n+1, generate a vector v uniformly distributed on the n-sphere, S^n, embed the n×n matrix in the next larger size with last column (0,...,0,1), and rotate the larger matrix so the last column becomes v.

As usual, we have special alternatives for the 3×3 case. Each of these methods begins with three independent random scalars uniformly distributed on the unit interval. Arvo (1992) takes advantage of the odd dimension to change a Householder reflection to a rotation by negation, and uses that to aim the axis of a uniform planar rotation.

Another method uses unit quaternions. Multiplication of rotation matrices is homomorphic to multiplication of quaternions, and multiplication by a unit quaternion rotates the unit sphere. Since the homomorphism is a local isometry, we immediately conclude that to produce a uniform distribution on $SO(3)$ we may use a uniform distribution on S^3.

Euler angles can also be used, though not with each angle uniformly distributed (Murnaghan 1962; Miles 1965).

For the axis-angle form, the axis is uniformly distributed over the unit sphere of directions, S^2, while the angle has the non-uniform distribution over $[0,\pi]$ noted previously (Miles 1965).

Aircraft Principal Axes

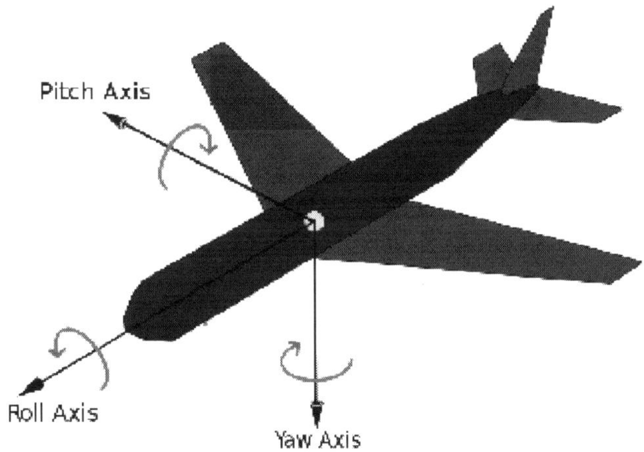

Figure: *The position of all three axes*

An aircraft in flight is free to rotate in three dimensions: *pitch*, nose up or down about an axis running from wing to wing), *yaw*, nose left or right about an axis running up and down; and *roll*, rotation about an axis running from nose to tail. The axes are alternatively designated as *lateral*, *vertical*, and *longitudinal*. These axes move with the vehicle, and rotate relative to the Earth along with the craft. These definitions were analogously applied to spacecraft when the first manned spacecraft were designed in the late 1950s.

These rotations are produced by torques (or moments) about the principal axes. On an aircraft, these are produced by means of moving control surfaces, which vary the distribution of the net aerodynamic force about the vehicle's centre of gravity. Elevators (moving flaps on the horizontal tail) produce pitch, a rudder on the vertical tail produces yaw, and ailerons (moving flaps on the wings) produce roll. On a spacecraft, the moments are usually produced by a reaction control system consisting of small rocket thrusters used to apply asymmetrical thrust on the vehicle.

Principal Axes

- *Vertical axis, or yaw axis:* an axis drawn from top to bottom, and perpendicular to the other two axes. Parallel to the *fuselage station.*
- *Lateral axis, transverse axis, or pitch axis:* an axis running from the pilot's left to right in piloted aircraft, and parallel to the wings of a winged aircraft. Parallel to the *buttock line.*
- *Longitudinal axis, or roll axis:* an axis drawn through the body of the vehicle from tail to nose in the normal direction of flight, or the direction the pilot faces. Parallel to the *waterline.*

Normally these axes are represented by the letters X, Y and Z in order to compare them with some reference frame, usually named x, y, z. Normally this is made in such a way that the X is used for the longitudinal axis, but there are other possibilities to do it.

Vertical Axis (Yaw)

Yaw axis is a vertical axis through an aircraft, rocket, or similar body, about which the body yaws; it may be a body, wind, or stability axis. Also known as yawing axis.

The yaw axis is defined to be perpendicular to the body of the wings with its origin at the centre of gravity and directed towards the bottom of the aircraft. A yaw motion is a movement of the nose of the aircraft from side to side. The pitch axis is perpendicular to the yaw axis and is parallel to the body of the wings with its origin at the centre of gravity and directed towards the right wing tip. A pitch motion is an up or down movement of the nose of the aircraft. The roll axis is perpendicular to the other two axes with its origin at the centre of gravity, and is directed towards the nose of the aircraft. A rolling motion is an up and down movement of the wing tips of the aircraft. The rudder is the primary control of yaw.

Lateral Axis (Pitch)

The lateral axis (also called transverse axis) passes through the plane from wingtip to wingtip. Rotation about this axis is called pitch. Pitch changes the vertical direction the aircraft's nose is pointing. The elevators are the primary control of pitch.

Longitudinal Axis (Roll)

The longitudinal axis passes through the plane from nose to tail. Rotation about this axis is called bank or roll. Bank changes the orientation of the aircraft's wings with respect to the downward force of gravity. The pilot changes bank angle by increasing the lift on one wing and decreasing it on the other. This differential lift causes bank rotation around the longitudinal axis. The ailerons are the primary control of bank. The rudder also has a secondary effect on bank.

Relationship with Other Systems of Axes

These axes are related to the principal axes of inertia, but are not the same. They are geometrical symmetry axes, regardless of the mass distribution of the aircraft. In aeronautical and aerospace engineering intrinsic rotations around these axes are often called Euler angles, but this conflicts with existing usage elsewhere. The calculus behind them is similar to the Frenet-Serret formulas. Performing a rotation in an intrinsic reference frame is equivalent to right-multiply its characteristic matrix (the matrix that has the vector of the reference frame as columns) by the matrix of the rotation.

History

The first aircraft to demonstrate active control about all three axes was the Wright brothers' 1902 glider.

Euler Angles

The Euler angles are three angles introduced by Leonhard Euler to describe the orientation of a rigid body. To describe such an orientation in 3-dimensional Euclidean space three parametres are required. They can be given in several ways, Euler angles being one of them. Euler angles also represent three composed rotations that move a reference frame to a given referred frame. This is equivalent to saying that any orientation can be achieved by composing three elemental rotations (rotations around a single axis), and also equivalent to saying that any rotation matrix can be decomposed as a product of three elemental rotation matrices. Without considering the possibilities of different signs for the angles or moving the reference frame, there are twelve different conventions divided in two groups. One of them is called "proper" Euler angles and the other Tait–Bryan angles. Sometimes "Euler angles" is used for all of them.

Proper Euler Angles

Sometimes Tait-Bryan angles are called Euler angles. When this is done, it is necessary to specify when classic Euler angles are referred, using an adjective like "proper" or similar.

Angles of a Given Frame: Geometric Derivation.

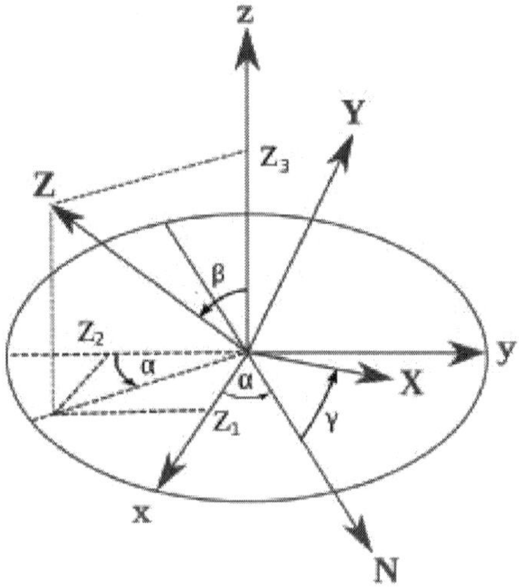

Figure: *Projections of Z vector.*

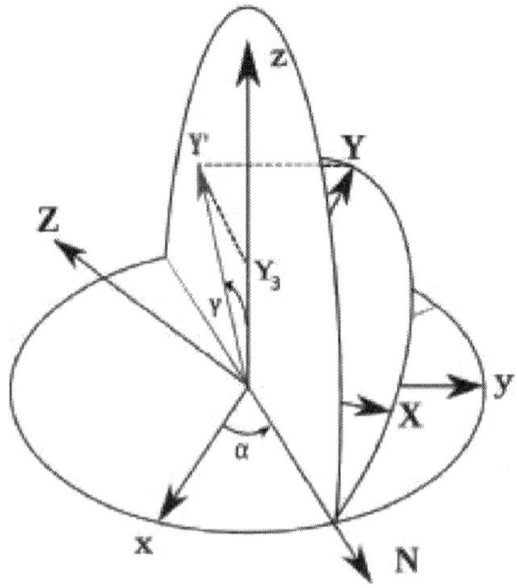

Figure: *Projections of Y vector.*

The fastest way to get the Euler Angles of a given frame is to write the three given vectors as columns of a matrix and compare it with the expression of the theoretical matrix. Hence the three Euler Angles can be calculated. Nevertheless, the same result can be reached avoiding matrix algebra, which is more geometrical. Assuming a frame with unitary vectors (*X*, *Y*, *Z*) as in the main diagram, it can be seen that

$$\cos(\beta) = Z_3$$

And, since

$$\sin^2 x = 1 - \cos^2 x$$

we have

$$\sin(\beta) = \sqrt{1 - Z_3^2}$$

As Z_2 is the double projection of a unitary vector,

$$\cos(\alpha)\cdot\sin(\beta) = -Z_2$$

$$\cos(\alpha) = -Z_2 / \sqrt{1 - Z_3^2}$$

There is a similar construction for Y_3, projecting it first over the plane defined by the axis z and the line of nodes. As the angle between the planes is $90 - \beta$ and $\cos(90 - \beta) = \sin(\beta)$, this leads to:

$$\sin(\beta)\cdot\cos(\gamma) = Y_3$$

$$\cos(\gamma) = Y_3 / \sqrt{1 - Z_3^2}$$

and finally, using the inverse cosine function,

$$\alpha = \arccos(-Z_2 / \sqrt{1 - Z_3^2})$$

$$\beta = \arccos(Z_3)$$

$$\gamma = \arccos(Y_3 / \sqrt{1 - Z_3^2})$$

It is interesting to note that the inverse cosine function yields two possible values for the argument. In this geometrical description only one of the solutions is valid. When Euler Angles are defined as a sequence of rotations all the solutions can be valid, but there will be only one inside the angle ranges. This is because the sequence of rotations to reach the target frame is not unique if the ranges are not previously defined. For computational purposes, it may be useful to represent the angles using atan2(*y*,*x*):

$$\alpha = \operatorname{atan2}(Z_1, -Z_2)$$

$$\gamma = \operatorname{atan2}(X_3, Y_3)$$

Properties

The Euler angles form a chart on all of $SO(3)$, the special orthogonal group of rotations in 3D space. The chart is smooth except for a polar coordinate style singularity along $\beta = 0$. The space of rotations is called in general "The Hypersphere of rotations", though this is a misnomer: the group Spin(3) is isometric to the hypersphere S^3, but the rotation space $SO(3)$ is instead isometric to the real projective space RP^3 which is a 2-fold quotient space of the hypersphere. This 2-to-1 ambiguity is the mathematical origin of spin in physics. A similar three angle decomposition applies to $SU(2)$, the special unitary group of rotations in complex 2D space, with the difference that β ranges from 0 to 2π. These are also called Euler angles.

The Haar measure for Euler angles has the simple form $\sin(\beta).d\alpha.d\beta.d\gamma$, usually normalised by a factor of $1/8\pi^2$. For example, to generate uniformly randomised orientations, let α and γ be uniform from 0 to 2π, let z be uniform from −1 to 1, and let $\beta = \arccos(z)$.

Applications

Vehicles and Moving Frames

Their main advantage over other orientation descriptions is that they are directly measurable from a gimbal mounted in a vehicle. As gyroscopes keep their rotation axis constant, angles measured in a gyro frame are equivalent to angles measured in the lab frame. Therefore gyros are used to know the actual orientation of moving spacecraft, and Euler angles are directly measurable. Intrinsic rotation angle cannot be read from a single gimbal, so there has to be more than one gimbal in a spacecraft. Normally there are at least three for redundancy. There is also a relation to the well-known gimbal lock problem of Mechanical Engineering. The most popular application is to describe aircraft attitudes, normally using a Tait–Bryan convention so that zero degrees elevation represents the horizontal attitude. Tait–Bryan angles represent the orientation of the aircraft respect a reference axis system (*world frame*) with three angles which in the context of an aircraft are normally called Heading, Elevation and Bank. When dealing with vehicles, different axes conventions are possible.

When studying rigid bodies in general, one calls the xyz system *space coordinates*, and the XYZ system *body coordinates*. The space coordinates are treated as unmoving, while the body coordinates are considered embedded in the moving body. Calculations involving acceleration, angular acceleration, angular velocity, angular momentum, and kinetic energy are often easiest in body coordinates, because then the moment of inertia tensor does not change in time. If one also diagonalises the rigid body's moment of inertia tensor (with nine components, six of which are independent), then one has a set of coordinates (called the principal axes) in which the moment of inertia tensor has only three components. The angular velocity of a rigid body takes a simple form using Euler angles in the moving frame. Also the Euler's rigid body equations are simpler because the inertia tensor is constant in that frame.

Axis-angle Representation

The axis-angle representation of a rotation, also known as the exponential coordinates of a rotation, parameterises a rotation by two values: a unit vector indicating the direction of a directed axis (straight line), and an angle describing the magnitude of the rotation about the axis. The rotation occurs in the sense prescribed by the right-hand rule. This representation evolves from Euler's rotation theorem, which implies that any rotation or sequence of rotations of a rigid body in a three-dimensional space is equivalent to a pure rotation about a single fixed axis. The axis-angle representation is equivalent to the more concise rotation vector, or Euler vector representation. In this case, both the axis and the angle are represented by a non-normalised vector codirectional with the axis whose magnitude is the rotation angle. Rodrigues' rotation formula can be used to apply to a vector a rotation represented by an axis and an angle.

Uses

The axis-angle representation is convenient when dealing with rigid body dynamics. It is useful to both characterise rotations, and also for converting between different representations of rigid body motion, such as homogeneous transformations and twists.

Example

Say you are standing on the ground and you pick the direction of gravity to be the negative z direction. Then if you turn to your left, you will travel $\frac{\pi}{2}$ radians (or 90 degrees) about the z axis. In axis-angle representation, this would be

$$\langle \text{axis}, \text{angle} \rangle = \left(\begin{bmatrix} a_x \\ a_y \\ a_z \end{bmatrix}, \theta \right) = \left(\begin{bmatrix} 0 \\ 0 \\ 1 \end{bmatrix}, \frac{\pi}{2} \right)$$

Rotation vector

The above example can be represented as a rotation vector with a magnitude of $\frac{\pi}{2}$ pointing in the z direction.

$$\begin{bmatrix} 0 \\ 0 \\ \frac{\pi}{2} \end{bmatrix}$$

This is the product of the angle and vector. It is more compact and is used for the exponential and log maps involving this representation.

Cylinder (Geometry)

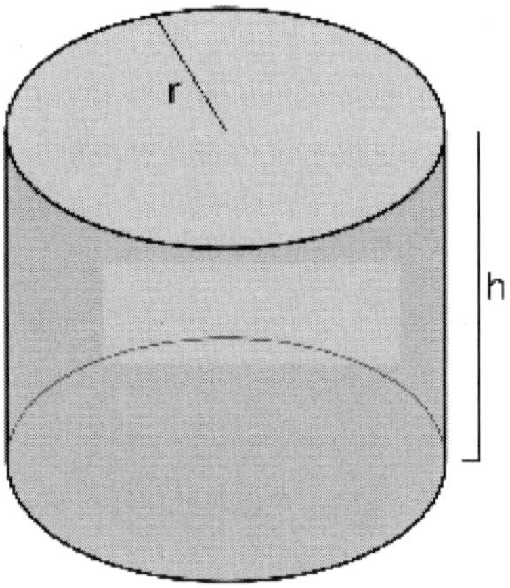

Figure: *A right circular cylinder with radius* r *and height* h.

A cylinder (from Greek – *kulindros*, "roller, tumbler") is one of the most basic curvilinear geometric shapes, the surface formed by the points at a fixed distance from a given line segment, the axis of the cylinder. The solid enclosed by this surface and by two planes perpendicular to the axis is also called a cylinder. The surface area and the volume of a cylinder have been known since deep antiquity. In differential geometry, a cylinder is defined more broadly as any ruled surface spanned by a one-parametre family of parallel lines. A cylinder whose cross section is an ellipse, parabola, or hyperbola is called an elliptic cylinder, parabolic cylinder, or hyperbolic cylinder respectively. The open cylinder is topologically equivalent to both the open annulus and the punctured plane.

Cylindric Section

Cylindric sections are the intersections of cylinders with planes. For a right circular cylinder, there are four possibilities. A plane tangent to the cylinder, meets the cylinder in a single straight line. Moved while parallel to itself, the plane either does not intersect the cylinder or intersects it in two parallel lines. All other planes intersect the cylinder in an ellipse or, when they are perpendicular to the axis of the cylinder, in a circle.

Other Types of Cylinders

An elliptic cylinder, or cylindroid, is a quadric surface, with the following equation in Cartesian coordinates:

$$\left(\frac{x}{a}\right)^2+\left(\frac{y}{b}\right)^2=1$$

This equation is for an elliptic cylinder, a generalisation of the ordinary, circular cylinder ($a = b$). Even more general is the generalised cylinder: the cross-section can be any curve. The cylinder is a *degenerate quadric* because at least one of the coordinates (in this case z) does not appear in the equation. An oblique cylinder has the top and bottom surfaces displaced from one another.

There are other more unusual types of cylinders. These are the *imaginary elliptic cylinders*:

$$\left(\frac{x}{a}\right)^2+\left(\frac{y}{b}\right)^2=-1$$

the *hyperbolic cylinder*:

$$\left(\frac{x}{a}\right)^2-\left(\frac{y}{b}\right)^2=1$$

and the *parabolic cylinder*:

$$x^2+2ay=0$$

About an Arbitrary Axis]

Consider an infinite cylinder whose axis lies along the vector

$$\vec{v}=(\alpha,\beta,\gamma)$$

We make use of spherical coordinates:

$$\rho^2=\alpha^2+\beta^2+\gamma^2$$

$$\theta = \arctan(\beta / \alpha)$$

$$\phi = \arcsin(\gamma / \rho)$$

These variables can be used to define A and B, the orthogonal vectors that form the basis for the cylinder:

$$A = -x\sin(\theta) + y\cos(\theta)cos(\phi) + z\cos(\theta)\sin(\phi)$$

$$B = -y\sin(\phi) + z\cos(\phi)$$

With these defined, we may use the familiar formula for a cylinder:

$$A^2 + B^2 = R^2$$

where R is the radius of the cylinder. These results are usually derived using rotation matrices.

Chapter

10

Surface Area of a Sphere

The surface area of a sphere is given by the following formula:

$$A = 4\pi r^2.$$

This formula was first derived by Archimedes, based upon the fact that the projection to the lateral surface of a circumscribed cylinder (i.e. the Lambert cylindrical equal-area projection) is area-preserving. It is also the derivative of the formula for the volume with respect to r because the total volume of a sphere of radius r can be thought of as the summation of the surface area of an infinite number of spherical shells of infinitesimal thickness concentrically stacked inside one another from radius 0 to radius r. At infinitesimal thickness the discrepancy between the inner and outer surface area of any given shell is infinitesimal and the elemental volume at radius r is simply the product of the surface area at radius r and the infinitesimal thickness.

At any given radius r, the incremental volume (δV) is given by the product of the surface area at radius r ($A(r)$) and the thickness of a shell (δr):

$$\delta V \approx A(r) \cdot \delta r.$$

The total volume is the summation of all shell volumes:

$$V \approx \sum A(r) \cdot \delta r.$$

In the limit as δr approaches zero this becomes:

$$V = \int_0^r A(r)dr.$$

Since we have already proved what the volume is, we can substitute V:

$$\frac{4}{3}\pi r^3 = \int_0^r A(r)dr.$$

Differentiating both sides of this equation with respect to r yields A as a function of r:

$$4\pi r^2 = A(r).$$

Which is generally abbreviated as:

$$A = 4\pi r^2.$$

Alternatively, the area element on the sphere is given in spherical coordinates by $dA = r^2 \sin\theta d\theta d\phi$. With Cartesian coordinates, the area element $dS = \frac{r}{\sqrt{r^2 - \sum_{i \neq k} x_i^2}} \Pi_{i \neq k} dx_i, \forall k$.

The total area can thus be obtained by integration:

$$A = \int_0^{2\pi} \int_0^{\pi} r^2 \sin\theta d\theta d\phi = 4\pi r^2.$$

Equations in R³

In analytic geometry, a sphere with centre (x_0, y_0, z_0) and radius r is the locus of all points (x, y, z) such that

$$(x - x_0)^2 + (y - y_0)^2 + (z - z_0)^2 = r^2.$$

The points on the sphere with radius r can be parameterised via

$$x = x_0 + r\cos\theta \sin\varphi$$

$$y = y_0 + r\sin\theta \sin\varphi \qquad (0 \leq \theta \leq 2\pi \text{ and } 0 \leq \varphi \leq \pi)$$

$$z = z_0 + r\cos\varphi$$

A sphere of any radius centreed at zero is an integral surface of the following differential form:

$$xdx + ydy + zdz = 0.$$

This equation reflects the fact that the position and velocity vectors of a point travelling on the sphere are always orthogonal to each other. The sphere has the smallest surface area among all surfaces enclosing a given volume and it encloses the largest volume among all closed surfaces with a given surface area. For this reason, the sphere appears in nature: for instance bubbles and small water drops are roughly spherical, because the surface tension locally minimises surface area. The surface area in relation to the mass of a sphere is called the specific surface area. From the above stated equations it can be expressed as follows:

$$SSA = \frac{A}{V\rho} = \frac{3}{r\rho}.$$

An image of one of the most accurate man-made spheres, as it refracts the image of Einstein in the background. This sphere was a fused quartz gyroscope for the Gravity Probe *B* experiment, and differs in shape from a perfect sphere by no more than 40 atoms of thickness.

It is thought that only non-rotating neutron stars are smoother. It was announced on 1 July 2008 that Australian scientists had created even more perfect spheres, accurate to 0.3 nanometres, as part of an international hunt to find a new global standard kilogram.

A sphere can also be defined as the surface formed by rotating a circle about any diametre. If the circle is replaced by an ellipse, and rotated about the major axis, the shape becomes a prolate spheroid, rotated about the minor axis, an oblate spheroid.

Terminology

Pairs of points on a sphere that lie on a straight line through its centre are called antipodal points. A great circle is a circle on the sphere that has the same centre and radius as the sphere, and consequently divides it into two equal parts. The shortest distance between two distinct non-antipodal points on the surface and measured along the surface, is on the unique great circle passing through the two points. Equipped with the great-circle distance, a great circle becomes the Riemannian circle.

If a particular point on a sphere is (arbitrarily) designated as its *north pole*, then the corresponding antipodal point is called the *south pole* and the equator is the great circle that is equidistant to them. Great circles through the two poles are called lines (or meridians) of longitude, and the line connecting the two poles is called the axis of rotation. Circles on the sphere that are parallel to the equator are lines of latitude. This terminology is also used for astronomical bodies such as the planet Earth, even though it is not spherical and only approximately spheroidal.

Hemisphere

A sphere is divided into two equal "hemispheres" by any plane that passes through its centre. If two intersecting planes pass through its centre, then they will subdivide the sphere into four lunes or biangles, the vertices of which all coincide with the antipodal points lying on the line of intersection of the planes.

The antipodal quotient of the sphere is the surface called the real projective plane, which can also be thought of as the northern hemisphere with antipodal points of the equator identified. The round hemisphere is conjectured to be the optimal (least area) filling of the Riemannian circle. If the planes don't pass through the sphere's centre, then the intersection is called *spheric section*.

N-Sphere

Just as a stereographic projection can project a sphere's surface to a plane, it can also project the surface of a 3-sphere into 3-space. This image shows three coordinate directions projected to 3-space: parallels (red), meridians (blue) and hypermeridians (green). Due to the conformal property of the stereographic projection, the curves intersect each other orthogonally (in the yellow points) as in 4D. All of the curves are circles: the curves that intersect <0,0,0,1> have an infinite radius (= straight line).

In mathematics, an n-sphere is a generalisation of the surface of an ordinary sphere to arbitrary dimension. For any natural number n, an n-sphere of radius r is defined as the set of points in $(n + 1)$-dimensional Euclidean space which are at distance r from a central point, where the radius r may be any positive real number. In symbols:

$$S^n\left\{x \in \mathbb{R}^{n+1} : \|x\| = r\right\}.$$

It is an n-dimensional manifold in Euclidean $(n + 1)$-space.

In particular:

a 0-sphere is a pair of points that are the ends of a line segment,

a 1-sphere is a circle in the plane, and

a 2-sphere is an ordinary sphere in three-dimensional space.

Spheres of dimension $n > 2$ are sometimes called hyperspheres, with 3-spheres sometimes known as glomes. The n-sphere of unit radius centreed at the origin is called the unit n-sphere, denoted S^n. The unit n-sphere is often referred to as *the* n-sphere. An n-sphere is the surface or boundary of an $(n + 1)$-dimensional ball, and is an

n-dimensional manifold. For $n \geq 2$, the n-spheres are the simply connected n-dimensional manifolds of constant, positive curvature. The n-spheres admit several other topological descriptions: for example, they can be constructed by gluing two n-dimensional Euclidean spaces together, by identifying the boundary of an n-cube with a point, or (inductively) by forming the suspension of an $(n - 1)$-sphere.

Description

For any natural number n, an n-sphere of radius r is defined as the set of points in $(n + 1)$-dimensional Euclidean space that are at distance r from some fixed point c, where r may be any positive real number and where c may be any point in $(n + 1)$-dimensional space. In particular:

- a 0-sphere is a pair of points $\{c - r, c + r\}$, and is the boundary of a line segment (1-ball).
- a 1-sphere is a circle of radius r centreed at c, and is the boundary of a disk (2-ball).
- a 2-sphere is an ordinary 2-dimensional sphere in 3-dimensional Euclidean space, and is the boundary of an ordinary ball (3-ball).
- a 3-sphere is a sphere in 4-dimensional Euclidean space.

Euclidean Coordinates in (n + 1)-space

The set of points in $(n + 1)$-space: $(x_1, x_1, x_2, \ldots, x_{n+1})$ that define an n-sphere, (S^n) is represented by the equation:

$$r^2 = \sum_{i=1}^{n+1} (x_i - c_i)^2.$$

where c is a centre point, and r is the radius.

The above n-sphere exists in $(n + 1)$-dimensional Euclidean space and is an example of an n-manifold. The volume form ω of an n-sphere of radius r is given by

$$\omega = \frac{1}{r} \sum_{j=1}^{n+1} (-1)^{j-1} x_j \, dx_1 \wedge \cdots \wedge dx_{j-1} \wedge dx_{j+1} \wedge \cdots \wedge dx_{n+1} = {*dr}$$

where * is the Hodge star operator. As a result, $dr \wedge \omega = dx_1 \wedge \cdots \wedge dx_{n+1}$.

n-*ball*

The space enclosed by an n-sphere is called an $(n + 1)$-ball. An $(n + 1)$-ball is closed if it includes the n-sphere, and it is open if it does not include the n-sphere.

Specifically:

- A 1-*ball*, a line segment, is the interior of a (0-sphere).
- A 2-*ball*, a disk, is the interior of a circle (1-sphere).
- A 3-*ball*, an ordinary ball, is the interior of a sphere (2-sphere).
- A 4-*ball*, is the interior of a 3-sphere, etc.

Topological Description

Topologically, an n-sphere can be constructed as a one-point compactification of n-dimensional Euclidean space. Briefly, the n-sphere can be described as $S^n = \mathbb{R}^n \cup \{\infty\}$, which is n-dimensional Euclidean space plus a single point representing infinity in all directions. In particular, if a single point is removed from an n-sphere, it becomes homeomorphic to $\mathbb{R}^n$. This forms the basis for stereographic projection.

Stereographic Projection

Just as a two dimensional sphere embedded in three dimensions can be mapped onto a two-dimensional plane by a stereographic projection, an n-sphere can be mapped onto an n-dimensional hyperplane by the n-dimensional version of the stereographic projection. For example, the point $[x,y,z]$ on a two-dimensional sphere of radius 1 maps to the point $\left[\frac{x}{1-z},\frac{y}{1-z}\right]$ on the xy plane. In other words,

$$[x,y,z] \mapsto \left[\frac{x}{1-z},\frac{y}{1-z}\right].$$

Likewise, the stereographic projection of an n-sphere $\mathbf{S}^{n-1}$ of radius 1 will map to the $n-1$ dimensional hyperplane $\mathbf{R}^{n-1}$ perpendicular to the x_n axis as

$$[x_1,x_2,\ldots,x_n] \mapsto \left[\frac{x_1}{1-x_n},\frac{x_2}{1-x_n},\ldots,\frac{x_{n-1}}{1-x_n}\right].$$

Generating Random Points

Uniformly at Random from the (n – 1)-sphere

To generate uniformly distributed random points on the $(n-1)$-sphere (*i.e.*, the surface of the n-ball), Marsaglia (1972) gives the following algorithm.

Generate an n-dimensional vector of normal deviates (it suffices to use N(0, 1), although in fact the choice of the variance is arbitrary),

$$\mathbf{x} = (x_1, x_2, \ldots, x_n)$$

Now calculate the "radius" of this point, $r = \sqrt{x_1^2 + x_2^2 + \cdots + x_n^2}$

The vector $\frac{1}{r}\mathbf{x}$ is uniformly distributed over the surface of the unit n-ball.

Examples

For example, when $n = 2$ the normal distribution $\exp(-x_1^2)$ when expanded over another axis $\exp(-x_2^2)$ after multiplication takes the form $\exp(-x_1^2 - x_2^2)$ or $\exp(-r^2)$ and so is only dependent on distance from the origin.

Alternatives

Another way to generate a random distribution on a hypersphere is to make a uniform distribution over a hypercube that includes the unit hyperball, exclude those points that are outside the hyperball, then project the remaining interior points outward from the origin onto the surface. This will give a uniform distribution, but it is necessary to remove the exterior points. As the relative volume of the hyperball to the hypercube decreases very rapidly with dimension, this procedure will succeed with high probability only for fairly small numbers of dimensions. Wendel's theorem gives the probability that all of the points generated will lie in the same half of the hypersphere.

Uniformly at Random from the n-ball

With a point selected from the surface of the n-ball uniformly at random, one needs only a radius to obtain a point uniformly at random within the n-ball. If u is a number generated uniformly at random from the interval [0, 1] and x is a point selected uniformly at random from the surface of the n-ball then $u^{1/n}x$ is uniformly distributed over the entire unit n-ball.

Specific Spheres

0-sphere: The pair of points $\{\pm R\}$ with the discrete topology for some $R > 0$. The only sphere that is disconnected. Has a natural Lie group structure; isomorphic to O(1). Parallelisable.

1-sphere: Also known as the circle. Has a nontrivial fundamental group. Abelian Lie group structure U(1); the circle group. Topologically equivalent to the real projective line, RP^1. Parallelisable. SO(2) = U(1).

2-sphere: Also known as the sphere. Equivalent to the complex projective line, CP^1. SO(3)/SO(2).

3-sphere: Parallelisable, Principal U(1)-bundle over the 2-sphere, Lie group structure Sp(1), where also

$$\mathrm{Sp}(1) \cong \mathrm{SO}(4) / \mathrm{SO}(3) \cong \mathrm{SU}(2) \cong \mathrm{Spin}(3)$$

4-sphere: Equivalent to the quaternionic projective line, HP^1. SO(5)/SO(4).

5-sphere: Principal U(1)-bundle over CP^2. SO(6)/SO(5) = SU(3)/SU(2).

6-sphere: Almost complex structure coming from the set of pure unit octonions. SO(7)/SO(6) = G_2/SU(3).

7-sphere: Topological quasigroup structure as the set of unit octonions. Principal Sp(1)-bundle over S^4. Parallelisable. SO(8)/SO(7) = SU(4)/SU(3) = Sp(2)/Sp(1) = Spin(7)/G_2 = Spin(6)/SU(3). The 7-sphere is of particular interest since it was in this dimension that the first exotic spheres were discovered.

8-sphere: Equivalent to the octonionic projective line OP^1.

23-sphere: A highly dense sphere-packing is possible in 24 dimensional space, which is related to the unique qualities of the Leech lattice.

Generalisation to Metric Spaces

More generally, in a metric space (E,d), the sphere of centre x and radius $r > 0$ is the set of points y such that $d(x,y) = r$.

If the centre is a distinguished point considered as origin of E, as in a normed space, it is not mentioned in the definition and notation. The same applies for the radius if it is taken to equal one, as in the case of a unit sphere. In contrast to a ball, a sphere may be an empty set, even for a large radius. For example, in Z^n with Euclidean metric, a sphere of radius r is nonempty only if r^2 can be written as sum of n squares of integers.

Topology

In topology, an n-sphere is defined as a space homeomorphic to the boundary of an $(n + 1)$-ball; thus, it is homeomorphic to the Euclidean n-sphere, but perhaps lacking its metric.

- a 0-sphere is a pair of points with the discrete topology
- a 1-sphere is a circle (up to homeomorphism); thus, for example, (the image of) any knot is a 1-sphere
- a 2-sphere is an ordinary sphere (up to homeomorphism); thus, for example, any spheroid is a 2-sphere

The n-sphere is denoted S^n. It is an example of a compact topological manifold without boundary. A sphere need not be smooth; if it is smooth, it need not be diffeomorphic to the Euclidean sphere.

The Heine–Borel theorem implies that a Euclidean n-sphere is compact. The sphere is the inverse image of a one-point set under the continuous function $\|x\|$. Therefore, the sphere is closed. S^n is also bounded; therefore it is compact.

Spherical Geometry

Spherical geometry is the geometry of the two-dimensional surface of a sphere. It is an example of a geometry which is not Euclidean. Two practical applications of the principles of spherical geometry are to navigation and astronomy.

In plane geometry the basic concepts are points and (straight) lines. On the sphere, points are defined in the usual sense.

The equivalents of lines are not defined in the usual sense of "straight line" in Euclidean geometry but in the sense of "the shortest paths between points," which are called geodesics. On the sphere the geodesics are the great circles; other geometric concepts are defined as in plane geometry but with straight lines replaced by great circles. Thus, in spherical geometry angles are defined between great circles, resulting in a spherical trigonometry that differs from ordinary trigonometry in many respects; for example, the sum of the interior angles of a triangle exceeds 180 degrees.

Spherical geometry is *not* elliptic geometry but shares with that geometry the property that a line has no parallels through a given point. Contrast this with Euclidean geometry, in which a line has one parallel through a given point, and hyperbolic geometry, in which a line has two parallels and an infinite number of ultraparallels through a given point.

An important geometry related to that of the sphere is that of the real projective plane; it is obtained by identifying antipodal points (pairs of opposite points) on the sphere. (This is elliptic geometry.) Locally, the projective plane has all the properties of spherical geometry, but it has different global properties. In particular, it is non-orientable, or one-sided.

Concepts of spherical geometry may also be applied to the oblong sphere, though minor modifications must be implemented on certain formulas. Higher-dimensional spherical geometries exist.

History

Spherical trigonometry was studied by early Greek mathematicians such as Menelaus of Alexandria, who wrote a book on spherical trigonometry called *Sphaerica* and developed Menelaus' theorem.

Islamic World

Muslims, according to Carra de Vaux, were "unquestionably the inventors of plane and spherical geometry, which did not, strictly speaking, exist among the Greeks". *The book of unknown arcs of a sphere* written by Islamic mathematician Al-Jayyani is considered to be the first treatise on spherical trigonometry. The book contains formulae for right-handed triangles, the general law of sines, and the solution of a spherical triangle by means of the polar triangle.

The book *On Triangles* by Regiomontanus, written around 1463, is the first pure trigonometrical work in Europe. However, Gerolamo Cardano noted a century later that much of the material there on spherical trigonometry was taken from the twelfth-century work of the Spanish Islamic scholar Jabir ibn Aflah.

Relation to Euclid's Postulates

Spherical geometry obeys two of Euclid's postulates: the second postulate ("to produce [extend] a finite straight line continuously in a straight line") and the fourth postulate ("that all right angles are equal to one another"). However, it violates the other three: contrary to the first postulate, there is not a unique shortest route between any two points (antipodal points such as the north and south poles on a spherical globe are counterexamples); contrary to the third postulate, a sphere does not contain circles of arbitrarily great radius; and contrary to the fifth (parallel) postulate, there is no point through which a line can be drawn that never intersects a given line.

A statement that is logically equivalent to the parallel postulate is that there exists a triangle whose angles add up to 180°. Since spherical geometry violates the parallel postulate, there exists no such triangle on the surface of a sphere. In fact, the sum of the angles of a triangle on a sphere is $180°(1+4f)$, where f is the fraction of the sphere's surface that is enclosed by the triangle. For any positive value of f, this exceeds 180°.

Eleven Properties of the Sphere

In their book *Geometry and the imagination* David Hilbert and Stephan Cohn-Vossen describe eleven properties of the sphere and discuss whether these properties uniquely determine the sphere. Several properties hold for the plane which can be thought of as a sphere with infinite radius. These properties are:

1. The points on the sphere are all the same distance from a fixed point. Also, the ratio of the distance of its points from two fixed points is constant.

The first part is the usual definition of the sphere and determines it uniquely. The second part can be easily deduced and follows a similar result of Apollonius of Perga for the circle. This second part also holds for the plane.

2. The contours and plane sections of the sphere are circles.

This property defines the sphere uniquely.

3. The sphere has constant width and constant girth.

The width of a surface is the distance between pairs of parallel tangent planes. There are numerous other closed convex surfaces which have constant width, for example the Meissner body. The girth of a surface is the circumference of the boundary of its orthogonal projection on to a plane. It can be proved that each of these properties implies the other.

1. *All points of a sphere are umbilics:* At any point on a surface we can find a normal direction which is at right angles to the surface, for the sphere these are the lines radiating out from the centre of the sphere. The intersection of a plane containing the normal with the surface will form a curve called a *normal section* and the curvature of this curve is the *normal curvature*. For most points on most surfaces, different sections will have different curvatures; the maximum and minimum values of these are called the principal curvatures. It can be proved that any closed surface will have at least four points called *umbilical points*. At an umbilic all the sectional curvatures are equal; in particular the principal curvatures are equal. Umbilical points can be thought of as the points where the surface is closely approximated by a sphere.

   For the sphere the curvatures of all normal sections are equal, so every point is an umbilic. The sphere and plane are the only surfaces with this property.

2. *The sphere does not have a surface of centres.* For a given normal section there is a circle whose curvature is the same

as the sectional curvature, is tangent to the surface and whose centre lines along on the normal line. Take the two centres corresponding to the maximum and minimum sectional curvatures: these are called the *focal points*, and the set of all such centres forms the focal surface.

For most surfaces the focal surface forms two sheets each of which is a surface and which come together at umbilical points. There are a number of special cases. For channel surfaces one sheet forms a curve and the other sheet is a surface; For cones, cylinders, toruses and cyclides both sheets form curves. For the sphere the centre of every osculating circle is at the centre of the sphere and the focal surface forms a single point. This is a unique property of the sphere.

3. *All geodesics of the sphere are closed curves.* Geodesics are curves on a surface which give the shortest distance between two points. They are a generalisation of the concept of a straight line in the plane. For the sphere the geodesics are great circles. There are many other surfaces with this property.
4. Of all the solids having a given volume, the sphere is the one with the smallest surface area; of all solids having a given surface area, the sphere is the one having the greatest volume.

These properties define the sphere uniquely. These properties can be seen by observing soap bubbles. A soap bubble will enclose a fixed volume and due to surface tension its surface area is minimal for that volume. This is why a free floating soap bubble approximates a sphere (though external forces such as gravity will distort the bubble's shape slightly).

5. *The sphere has the smallest total mean curvature among all convex solids with a given surface area.* The mean curvature is the average of the two principal curvatures and as these are constant at all points of the sphere then so is the mean curvature.
6. *The sphere has constant mean curvature.* The sphere is the only imbedded surface without boundary or singularities with constant positive mean curvature. There are other immersed surfaces with constant mean curvature. The minimal surfaces have zero mean curvature.
7. *The sphere has constant positive Gaussian curvature.* Gaussian curvature is the product of the two principal curvatures. It is an intrinsic property which can be determined by measuring

length and angles and does not depend on the way the surface is embedded in space. Hence, bending a surface will not alter the Gaussian curvature and other surfaces with constant positive Gaussian curvature can be obtained by cutting a small slit in the sphere and bending it. All these other surfaces would have boundaries and the sphere is the only surface without boundary with constant positive Gaussian curvature. The pseudosphere is an example of a surface with constant negative Gaussian curvature.

8. *The sphere is transformed into itself by a three-parametre family of rigid motions.* Consider a unit sphere placed at the origin, a rotation around the x, y or z axis will map the sphere onto itself, indeed any rotation about a line through the origin can be expressed as a combination of rotations around the three-coordinate axis. Thus there is a three-parametre family of rotations which transform the sphere onto itself, this is the rotation group $SO(3)$. The plane is the only other surface with a three-parametre family of transformations (translations along the x and y axis and rotations around the origin). Circular cylinders are the only surfaces with two-parametre families of rigid motions and the surfaces of revolution and helicoids are the only surfaces with a one-parametre family.

Cubes in Relation to Spheres

For every sphere there are multiple cuboids that may be inscribed within the sphere. The largest cuboid which can be inscribed within a sphere is a cube.

Degree (Angle)

A degree (in full, a degree of arc, arc degree, or arcdegree), usually denoted by ° (the degree symbol), is a measurement of plane angle, representing $^{1}D_{360}$ of a full rotation; one degree is equivalent to $\pi/180$ radians. It is not an *SI* unit, as the *SI* unit for angles is radian, but it is mentioned in the *SI* brochure as an accepted unit.

History

The original motivation for choosing the degree as a unit of rotations and angles is unknown. One theory states that it is related to the fact that 360 is approximately the number of days in a year. Ancient astronomers noticed that the sun, that follows through the ecliptic path over the course of the year, seems to advance in that path by approximately one degree, each day. Some ancient calendars, such as the Persian calendar, used 360 days for a year. The Mayans used 20 cycles of 18 plus 5 unlucky days in one of their Maya calendars. The use of a calendar with 360 days may be related to the use of sexagesimal numbers.

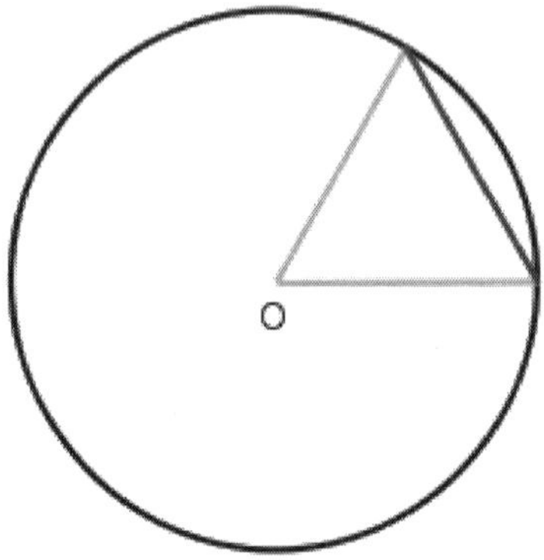

Figure: *A circle with an equilateral chord (gey). One sixtieth of this arc is a degree. Six such chords complete the circle.*

Another theory is that the Babylonians subdivided the circle using the angle of an equilateral triangle as the basic unit and further subdivided the latter into 60 parts following their sexagesimal numeric system. The earliest trigonometry, used by the Babylonian astronomers and their Greek successors, was based on chords of a circle. A chord of length equal to the radius made a natural base quantity. One sixtieth of this, using their standard sexagesimal divisions, was a degree.

Aristarchus of Samos and Hipparchos seem to have been among the first Greek scientists to exploit Babylonian astronomical knowledge and techniques systematically. Timocharis, Aristarchus, Aristillus, Archimedes, and Hipparchus were the first Greeks known to divide the circle in 360 degrees of 60 arc minutes (DIO 14 ‡2 p.19 n.24). Eratosthenes used a simpler sexagesimal system dividing a circle into 60 parts. The division of the circle into 360 parts also occurred in ancient India, as evidenced in the Rigveda:

Twelve spokes, one wheel, navels three.

Who can comprehend this?

On it are placed together

three hundred and sixty like pegs.

They shake not in the least.

— *Dirghatamas , Rigveda 1.164.48*

Another motivation for choosing the number 360 may have been that it is readily divisible: 360 has 24 divisors, making it one of only 7 numbers that have more divisors than any number twice itself (sequence A072938 in OEIS).

Furthermore, it is divisable by every number from 1 to 10 except 7. This property has many useful applications, such as dividing the world into 24 time zones, each of which is nominally 15° of longitude, to correlate with the established 24-hour day convention.

Finally, it may be the case that more than one of these factors has come into play.

According to that theory, the number is approximately 365 because of the apparent movement of the sun against the celestial sphere and that it was rounded to 360 for some of the mathematical reasons cited above.

Subdivisions

For many practical purposes, a degree is a small enough angle that whole degrees provide sufficient precision.

When this is not the case, as in astronomy or for latitudes and longitudes on the Earth, degree measurements may be written with decimal places like 40.1875° with the degree symbol behind the decimals.

Alternatively, the traditional sexagesimal unit subdivision can be used. One degree is divided into 60 *minutes (of arc)*, and one minute into 60 *seconds (of arc)*.

These units, also called the *arcminute* and *arcsecond*, are respectively represented as a single and double prime: for example, 40.1875° = 40° 112 153 . Sometimes single and double quotation marks are used instead: 40° 11' 15– .

If still more accuracy is required, current practice is to use decimal divisions of the second like 40° 112 15.43 . The older system of thirds, fourths, etc., which continues the sexagesimal unit subdivision, was used by al-Kashi and other ancient astronomers, but is rarely used today.

These subdivisions were denoted by writing the Roman numeral for the number of sixtieths in superscript: 1 for a "prime" (minute of arc), 1^{II} for a second, 1^{III} for a third, 1^{IV} for a fourth, etc. Hence the modern symbols for the minute and second of arc, and the word "second" also refer to this system.

Alternative Units

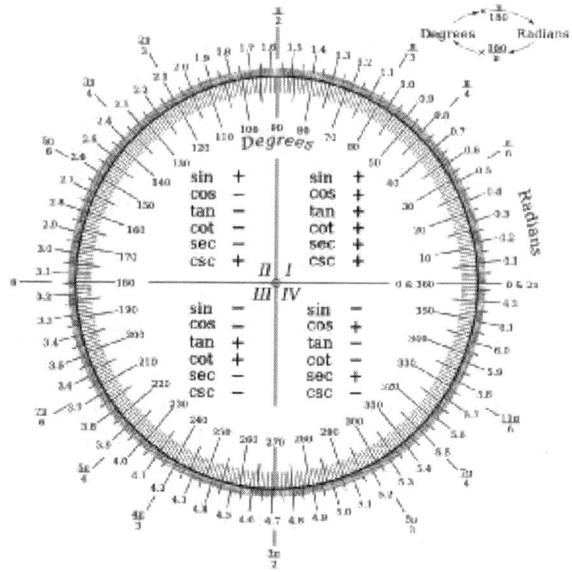

Figure: *A chart to convert between degrees and radians*

In most mathematical work beyond practical geometry, angles are typically measured in radians rather than degrees. This is for a variety of reasons; for example, the trigonometric functions have simpler and more "natural" properties when their arguments are expressed in radians. These considerations outweigh the convenient divisibility of the number 360. One complete turn (360°) is equal to 2π radians, so 180° is equal to π radians, or equivalently, the degree is a mathematical constant: $1° = {}^{\pi}D_{180}$. The turn (or revolution, full ;circle, full rotation, cycle) is used in technology and science. 1 turn = 360°.

With the invention of the metric system, based on powers of ten, there was an attempt to define a "decimal degree" (grad or gon), so that the number of decimal degrees in a right angle would be 100 *gon*, and there would be 400 *gon* in a circle. Although this idea was abandoned already by Napoleon, some groups have continued to use it and many scientific calculators still support it.

An angular mil, which is most used in military applications, has at least three specific variants, ranging from ${}^{1}D_{6400}$ to ${}^{1}D_{6000}$, each approximately equal to one milliradian. However, ${}^{1}D_{6000}$ used by the Russian Army originated in Imperial Russia, where an equilateral chord was divided into tenths to give a circle of 600 units (this may be seen on a protractor, circa 1900, in the St Petersberg Museum of Artillery).

Angle

In geometry, an angle is the figure formed by two rays, called the sides of the angle, sharing a common endpoint, called the vertex of the angle. Angles are usually presumed to be in a Euclidean plane, but are also defined in non-Euclidean geometry.

Angle is also used to designate the measure of an angle or of a rotation. This measure is the ratio of the length of a circular arc by its radius.

In the case of an angle (figure), the arc is centreed at the vertex and delimited by the sides. In the case of a rotation, the arc is centreed at the centre of the rotation and delimited by any point and its image by the rotation.

The word *angle* comes from the Latin word *angulus*, meaning "a corner". The word *angulus* is a diminutive, of which the primitive form, *angus*, does not occur in Latin. Cognate words are the Greek ãêýëïò *(ankylïs)*, meaning "crooked, curved," and the English word "ankle". Both are connected with the Proto-Indo-European root **ank-*, meaning "to bend" or "bow".

Euclid defines a plane angle as the inclination to each other, in a plane, of two lines which meet each other, and do not lie straight with respect to each other. According to Proclus an angle must be either a quality or a quantity, or a relationship. The first concept was used by Eudemus, who regarded an angle as a deviation from a straight line; the second by Carpus of Antioch, who regarded it as the interval or space between the intersecting lines; Euclid adopted the third concept, although his definitions of right, acute, and obtuse angles are certainly quantitative.

Measuring Angles

The size of an angle is characterised by the magnitude of the smallest rotation that maps one of the rays into the other. Angles that have the same size are called congruent angles.

In some contexts, such as identifying a point on a circle or describing the *orientation* of an object in two dimensions relative to a reference orientation, angles that differ by an exact multiple of a full turn are effectively equivalent. In other contexts, such as identifying a point on a spiral curve or describing the *cumulative rotation* of an object in two dimensions relative to a reference orientation, angles that differ by a non-zero multiple of a full turn are not equivalent.

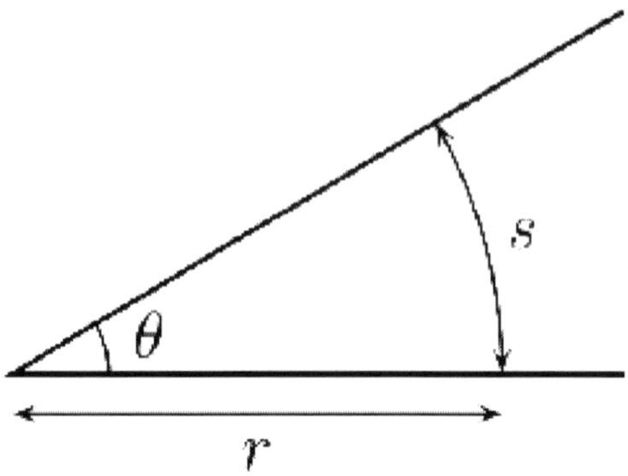

Figure: *The measure of angle θ is the quotient of s and r.*

In order to measure an angle θ, a circular arc centreed at the vertex of the angle is drawn, e.g. with a pair of compasses. The length of the arc *s* is then divided by the radius of the arc *r*, and possibly multiplied by a scaling constant *k* (which depends on the units of measurement that are chosen):

$$\theta = k\frac{s}{r}.$$

The value of θ thus defined is independent of the size of the circle: if the length of the radius is changed then the arc length changes in the same proportion, so the ratio *s/r* is unaltered.

Units

Units used to represent angles are listed below in descending magnitude order. Of these units, the *degree* and the *radian* are by far the most commonly used. Angles expressed in radians are dimensionless for the purposes of dimensional analysis. Most units of angular measurement are defined such that one *turn* (i.e. one full circle) is equal to *n* units, for some whole number *n*. The two exceptions are the radian and the diametre part. For example, in the case of degrees, $n = 360$. A *turn* of *n* units is obtained by setting $k = n/(2\pi)$ in the formula above. (Proof. The formula above can be rewritten as $k = \theta r/s$. One turn, for which $\theta = n$ units, corresponds to an arc equal in length to the circle's circumference, which is $2\pi r$, so $s = 2\pi r$. Substituting *n* for θ and $2\pi r$ for *s* in the formula, results in $k = nr/(2\pi r) = n/(2\pi)$.)

- The turn (or full circle, revolution, rotation, or cycle) is one full circle. A *turn* can be subdivided in centiturns and milliturns. A turn is abbreviated τ or *rev* or *rot* depending on the application, but just *r* in *rpm* (revolutions per minute). 1 *turn* = 360° = 2π rad = 400 grad = 4 right angles.

- The *quadrant* is 1/4 of a turn, i.e. a right angle. It is the unit used in Euclid's Elements. 1 quad. = 90° = $\pi/2$ rad = 1/4 turn = 100 grad. In German the symbol “ has been used to denote a quadrant.
- The angle of the equilateral triangle is 1/6 of a turn. It was the unit used by the Babylonians, and is especially easy to construct with ruler and compasses. The degree, minute of arc and second of arc are sexagesimal subunits of the Babylonian unit. 1 Babylonian unit = 60° = $\pi/3$ rad ≈ 1.047197551 rad.

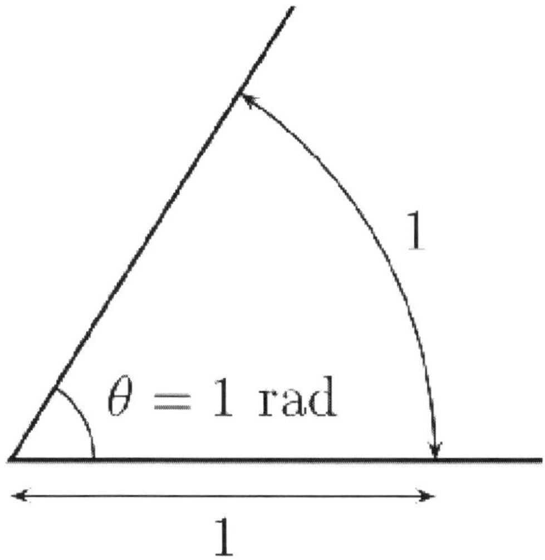

Figure: *θ = s/r rad = 1 rad.*

- The radian is the angle subtended by an arc of a circle that has the same length as the circle's radius (k = 1 in the formula given earlier). One turn is 2π radians, and one radian is $180/\pi$ degrees, or about 57.2958 degrees. The radian is abbreviated *rad*, though this symbol is often omitted in mathematical texts, where radians are assumed unless specified otherwise. When radians are used angles are considered as dimensionless. The radian is used in virtually all mathematical work beyond simple practical geometry, due, for example, to the pleasing and “natural” properties that the trigonometric functions display when their arguments are in radians. The radian is the (derived) unit of angular measurement in the SI system.
- The diametre part (occasionally used in Islamic mathematics) is 1/60 radian. One “diametre part” is approximately 0.95493°.

- The astronomical hour angle is 1/24 of a turn. Since this system is amenable to measuring objects that cycle once per day (such as the relative position of stars), the sexagesimal subunits are called minute of time and second of time. Note that these are distinct from, and 15 times larger than, minutes and seconds of arc. 1 hour = 15° = $\pi/12$ rad = 1/6 quad. = 1/24 *turn* ≈ 16.667 grad.
- The point, used in navigation, is 1/32 of a turn. 1 point = 1/8 of a right angle = 11.25° = 12.5 grad. Each point is subdivided in four quarter-points so that 1 turn equals 128 quarter-points.
- Eratosthenes used a unit of 6° so that a whole turn was divided in 60 units.
- The Babylonians sometimes used the unit pechus of about 2° or 2½°.
- The binary degree, also known as the binary radian (or brad), is 1/256 of a turn. The binary degree is used in computing so that an angle can be efficiently represented in a single byte (albeit to limited precision). Other measures of angle used in computing may be based on dividing one whole turn into 2 equal parts for other values of *n*.
- The degree, denoted by a small superscript circle (°), is 1/360 of a turn, so one *turn* is 360°. One advantage of this old sexagesimal subunit is that many angles common in simple geometry are measured as a whole number of degrees. Fractions of a degree may be written in normal decimal notation (e.g. 3.5° for three and a half degrees), but the "minute" and "second" sexagesimal subunits of the "degree-minute-second" system are also in use, especially for geographical coordinates and in astronomy and ballistics:
- The grad, also called grade, gradian, or gon, is 1/400 of a turn, so a right angle is 100 grads. It is a decimal subunit of the quadrant. A kilometre was historically defined as a centi-grad of arc along a great circle of the Earth, so the kilometre is the decimal analog to the sexagesimal nautical mile. The grad is used mostly in triangulation.
- The mil is *approximately* equal to a milliradian. There are several definitions ranging from 0.05625 to 0.06 degrees (3.375 to 3.6 minutes), with the milliradian being approximately

0.05729578 degrees (3.43775 minutes). In NATO countries, it is defined as 1/6400th of a circle. Its value is approximately equal to the angle subtended by a width of 1 metre as seen from 1 km away (2π / 6400 = 0.0009817... R" 1/1000).

- The minute of arc (or MOA, arcminute, or just minute) is 1/60 of a degree = 1/21600 turn. It is denoted by a single prime (2). For example, 3° 302 is equal to 3 + 30/60 degrees, or 3.5 degrees. A mixed format with decimal fractions is also sometimes used, e.g. 3° 5.722 = 3 + 5.72/60 degrees. A nautical mile was historically defined as a minute of arc along a great circle of the Earth.
- The second of arc (or arcsecond, or just second) is 1/60 of a minute of arc and 1/3600 of a degree. It is denoted by a double prime (3). For example, 3° 72 303 is equal to 3 + 7/60 + 30/3600 degrees, or 3.125 degrees.

Positive and Negative Angles

Although the definition of the measurement of an angle does not support the concept of a negative angle, it is frequently useful to impose a convention that allows positive and negative angular values to represent orientations and/or rotations in opposite directions relative to some reference.

In a two dimensional Cartesian coordinate system, angles are typically defined relative to the positive x-axis with positive angles representing rotations towards the positive y-axis and negative angles representing rotations towards the negative y-axis. When Cartesian coordinates are represented as they commonly are, with the x-axis rightward and the y-axis upward, positive rotations are counterclockwise and negative rotations are clockwise.

In many contexts, an angle of $-\theta$ is effectively equivalent to an angle of "one full turn minus θ". For example, an orientation represented as – 45° is effectively equivalent to an orientation represented as 360° – 45° or 315°. However, a rotation of – 45° would not be the same as a rotation of 315°.

In three dimensional geometry, "clockwise" and "counterclockwise" have no absolute meaning, so the direction of positive and negative angles must be defined relative to some reference, which is typically a vector passing through the angle's vertex and perpendicular to the plane in which the rays of the angle lie.

In navigation, bearings are measured relative to north. By convention, viewed from above, bearing angle are positive clockwise, so a bearing of 45° corresponds to a north-east orientation. Negative bearings are not used in navigation, so a north-west orientation corresponds to a bearing of 315°.

Alternative Ways of Measuring the Size of an Angle

There are several alternatives to measuring the size of an angle by the corresponding angle of rotation. The grade of a slope, or gradient is equal to the tangent of the angle, or sometimes the sine. Gradients are often expressed as a percentage. For very small values (less than 5%), the grade of a slope is approximately the measure of an angle in radians.

In rational geometry the *spread* between two lines is defined at the square of sine of the angle between the lines. Since the sine of an angle and the sine of its supplementary angle are the same any angle of rotation that maps one of the lines into the other leads to the same value of the spread between the lines.

Astronomical Approximations

Astronomers measure angular separation of objects in degrees from their point of observation.

- 1° is approximately the width of a little finger at arm's length.
- 10° is approximately the width of a closed fist at arm's length.
- 20° is approximately the width of a handspan at arm's length.

These measurements clearly depend on the individual subject, and the above should be treated as rough approximations only.

Identifying Angles

In mathematical expressions, it is common to use Greek letters (α, β, γ, θ, φ, ...) to serve as variables standing for the size of some angle. (To avoid confusion with its other meaning, the symbol π is typically not used for this purpose.) Lower case roman letters (a, b, c, ...) are also used.

In geometric figures, angles may also be identified by the labels attached to the three points that define them. For example, the angle at vertex A enclosed by the rays AB and AC (i.e. the lines from point A to point B and point A to point C) is denoted "BAC or $B\hat{A}C$. Sometimes, where there is no risk of confusion, the angle may be referred to simply by its vertex ("angle A").

Potentially, an angle denoted, say, "*BAC* might refer to any of four angles: the clockwise angle from *B* to *C*, the anticlockwise angle from *B* to *C*, the clockwise angle from *C* to *B*, or the anticlockwise angle from *C* to *B*, where the direction in which the angle is measured determines its sign.

However, in many geometrical situations it is obvious from context that the positive angle less than or equal to 180 degrees is meant, and no ambiguity arises. Otherwise, a convention may be adopted so that "*BAC* always refers to the anticlockwise (positive) angle from *B* to *C*, and "*CAB* to the anticlockwise (positive) angle from *C* to *B*.

Types of Angles

- An angle equal to 1/4 turn (90° or π/2 radians) is called a right angle.

  Two lines that form a right angle are said to be perpendicular or orthogonal.
- Angles equal to 1/2 turn (180° or two right angles) are called straight angles.
- Angles equal to 1 turn (360° or four right angles) are called full angles.
- Angles that are not right angles or a multiple of a right angle are called oblique angles.
- Angles smaller than a right angle (less than 90°) are called acute angles ("acute" meaning "sharp").
- Angles larger than a right angle and smaller than a straight angle (between 90° and 180°) are called obtuse angles ("obtuse" meaning "blunt").
- Angles larger than a straight angle but less than 1 turn (between 180° and 360°) are called reflex angles.
- Angles that have the same measure (i.e. the same magnitude) are said to be equal *(UK)* or congruent *(USA)*. An angle is defined by its measure and is not dependent upon the lengths of the sides of the angle (e.g. all *right angles* are *congruent*).
- Two angles opposite each other, formed by two intersecting straight lines that form an "*X*"-like shape, are called vertical angles or opposite angles or vertically opposite angles. These angles are equal in measure.

- Angles that share a common vertex and edge but do not share any interior points are called adjacent angles.
- Two angles that sum to one right angle (90°) are called complementary angles.

  The difference between an angle and a right angle is termed the complement of the angle.
- Two angles that sum to a straight angle (180°) are called supplementary angles.

  The difference between an angle and a straight angle (180°) is termed the supplement of the angle.
- Two angles that sum to one turn (360°) are called explementary angles or conjugate angles.
- An angle that is part of a simple polygon is called an interior angle if it lies on the inside of that simple polygon. A concave simple polygon has at least one interior angle that exceeds 180°.

In Euclidean geometry, the measures of the interior angles of a triangle add up to π radians, or 180°, or 1/2 turn; the measures of the interior angles of a simple quadrilateral add up to 2π radians, or 360°, or 1 turn. In general, the measures of the interior angles of a simple polygon with n sides add up to $[(n - 2) \times \pi]$ radians, or $[(n - 2) \times 180]°$, or $(2n - 4)$ right angles, or $(n/2 - 1)$ turn.

The angle supplementary to the interior angle is called the exterior angle. It measures the amount of rotation one has to make at this vertex to trace out the polygon.

If the corresponding interior angle is a reflex angle, the exterior angle should be considered negative. Even in a non-simple polygon it may be possible to define the exterior angle, but one will have to pick an orientation of the plane (or surface) to decide the sign of the exterior angle measure.

In Euclidean geometry, the sum of the exterior angles of a simple polygon will be one full turn (360°).

- Some authors use the name exterior angle of a simple polygon to simply mean the explementary (*not* supplementary!) of the interior angle. This conflicts with the above usage.
- The angle between two planes (such as two adjacent faces of a polyhedron) is called a dihedral angle. It may be defined as the acute angle between two lines normal to the planes.

- The angle between a plane and an intersecting straight line is equal to ninety degrees minus the angle between the intersecting line and the line that goes through the point of intersection and is normal to the plane.
- Alternate angles, corresponding angle, interior angles and exterior angles are associated with a transversal of a pair of lines by a third.
- A reference angle is the acute version of any angle determined by repeatedly subtracting or adding 180 degrees, and subtracting the result from 180 degrees if necessary, until a value between 0 degrees and 90 degrees is obtained. For example, an angle of 30 degrees has a reference angle of 30 degrees, and an angle of 150 degrees also has a reference angle of 30 degrees (180-150). An angle of 750 degrees has a reference angle of 30 degrees (750-720).

Angles Between Curves

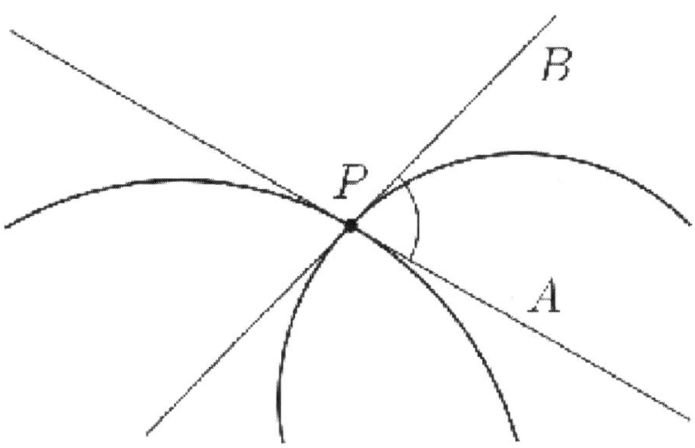

Figure: *The angle between the two curves at P is defined as the angle between the tangents A and B at P*

The angle between a line and a curve (mixed angle) or between two intersecting curves (curvilinear angle) is defined to be the angle between the tangents at the point of intersection. Various names (now rarely, if ever, used) have been given to particular cases:—*amphicyrtic* or *cissoidal* biconvex; *xystroidal* or *sistroidal*, concavo-convex; *amphicoelic* or *angulus lunularis*, biconcave.

Chapter

11

Dot Product and Generalisation

In the Euclidean plane, the angle θ between two vectors u and v is related to their dot product and their lengths by the formula

$$\mathbf{u}\cdot\mathbf{v} = \cos(\theta)\|\mathbf{u}\|\|\mathbf{v}\|$$

This formula supplies an easy method to find the angle between two planes (or curved surfaces) from their normal vectors and between skew lines from their vector equations.

Angles Between Subspaces

The definition of the angle between one-dimensional subspaces $\text{span}(\mathbf{u})$ and $\text{span}(\mathbf{v})$ given by

$$|<\mathbf{u},\mathbf{v}>| = \cos(\theta)\|\mathbf{u}\|\|\mathbf{v}\|$$

in a Hilbert space can be extended to subspaces of any finite dimensions. Given two subspaces $\mathcal{U},\mathcal{W}$ with $\dim(\mathcal{U}) := k \le \dim(\mathcal{W}) := l$ this leads to a definition of k angles called canonical or principal angles between subspaces.

Angles in Riemannian Geometry

In Riemannian geometry, the metric tensor is used to define the angle between two tangents. Where U and V are tangent vectors and g_{ij} are the components of the metric tensor G,

$$\cos\theta = \frac{g_{ij}U^iV^j}{\sqrt{\left|g_{ij}U^iU^j\right|\left|g_{ij}V^iV^j\right|}}.$$

Angles in Geography and Astronomy

In geography, the location of any point on the Earth can be identified using a geographic coordinate system. This system specifies the latitude and longitude of any location in terms of angles subtended at the centre of the Earth, using the equator and (usually) the Greenwich meridian as references.

In astronomy, a given point on the celestial sphere (that is, the apparent position of an astronomical object) can be identified using any of several astronomical coordinate systems, where the references vary according to the particular system. Astronomers measure the angular separation of two stars by imagining two lines through the centre of the Earth, each intersecting one of the stars. The angle between those lines can be measured, and is the angular separation between the two stars.

Astronomers also measure the apparent size of objects as an angular diametre. For example, the full moon has an angular diametre of approximately 0.5°, when viewed from Earth. One could say, "The Moon's diametre subtends an angle of half a degree." The small-angle formula can be used to convert such an angular measurement into a distance/size ratio.

Cone (Geometry)

A cone is an n-dimensional geometric shape that tapers smoothly from a base (usually flat and circular) to a point called the apex or vertex.

Formally, it is the solid figure formed by the locus of all straight line segments that join the apex to the base. The term "cone" is sometimes used to refer to the surface or the lateral surface of this solid figure (the lateral surface of a cone is equal to the surface minus the base).

The axis of a cone is the straight line (if any), passing through the apex, about which the base has a rotational symmetry.

In common usage in elementary geometry, cones are assumed to be right circular, where *right* means that the axis passes through the centre of the base (suitably defined) at right angles to its plane, and *circular* means that the base is a circle. Contrasted with right cones are oblique cones, in which the axis does not pass perpendicularly through the centre of the base. In general, however, the base may be any shape, and the apex may lie anywhere (though it is often assumed that the base is bounded and has finite area, and that the apex lies outside the plane of the base). For example, a pyramid is technically a cone with a polygonal base.

Other Mathematical Meanings

In mathematical usage, the word "cone" is used also for an 'infinite cone', the union of a set of half-lines that start at a common apex point and go through a base. Observe that an infinite cone is not bounded by its base and extends to infinity. A 'doubly infinite cone', or 'double cone', is the union of a set of straight lines that pass through a common apex point and go through a base, therefore double infinite cones extend symmetrically on both sides of the apex.

The boundary of an infinite or doubly infinite cone is a conical surface, and the intersection of a plane with this surface is a conic section. For infinite cones, the word *axis* again usually refers to the axis of rotational symmetry (if any). Either half of a double cone on one side of the apex is called a 'nappe'. Depending on the context, "cone" may also mean specifically a convex cone or a projective cone.

Also, an oblique shaped cone is known as a circular based pyramid due to the inwardly converging hypos towards the apex point.

Further Terminology

The perimetre of the base of a cone is called the 'directrix', and each of the line segments between the directrix and apex is a 'generatrix' of the lateral surface. The 'base radius' of a circular cone is the radius of its base; often this is simply called the radius of the cone. The aperture of a right circular cone is the maximum angle between two generatrix lines; if the generatrix makes an angle θ to the axis, the aperture is 2θ.

A cone with its apex cut off by a plane is called a "truncated cone"; if the truncation plane is parallel to the cone's base, it is called a frustum. An 'elliptical cone' is a cone with an elliptical base. A 'generalised cone' is the surface created by the set of lines passing through a vertex and every point on a boundary.

Conical Surface

In geometry, a (general) conical surface is the unbounded surface formed by the union of all the straight lines that pass through a fixed point — the *apex* or *vertex* — and any point of some fixed space curve — the *directrix* — that does not contain the apex. Each of those lines is called a *generatrix* of the surface.

Every conic surface is ruled and developable. In general, a conical surface consists of two congruent unbounded halves joined by the apex. Each half is called a nappe, and is the union of all the rays that start at the apex and pass through a point of some fixed space curve. (In some cases, however, the two nappes may intersect, or even coincide with the full surface.) Sometimes the term "conical surface" is used to mean just one nappe.

If the directrix is a circle C, and the apex is located on the circle's *axis* (the line that contains the centre of and is perpendicular to its plane), one obtains the *right circular conical surface*. This special case is often called a *cone*, because it is one of the two distinct surfaces that bound the geometric solid of that name. This geometric object can also be described as the set of all points swept by a line that intercepts the axis and rotates around it; or the union of all lines that intersect the axis at a fixed point and at a fixed angle. The *aperture* of the cone is the angle.

More generally, when the directrix is an ellipse, or any conic section, and the apex is an arbitrary point not on the plane of , one obtains a *conical quadric*, which is a special case of a quadric surface.

A cylindrical surface can be viewed as a limiting case of a conical surface whose apex is moved off to infinity in a particular direction. Indeed, in projective geometry a cylindrical surface is just a special case of a conical surface.

Surface of Constant Width

In geometry, a surface of constant width is a convex form whose width, measured by the distance between two opposite parallel planes touching its boundary, is the same regardless of the direction of those two parallel planes. One defines the width of the surface in a given direction to be the perpendicular distance between the parallels perpendicular to that direction. Thus, a surface of constant width is the three-dimensional analogue of a curve of constant width, a two-dimensional shape with a constant distance between pairs of parallel tangent lines.

More generally, any compact convex body D has one pair of parallel supporting planes in a given direction. A supporting plane is a plane that intersects the boundary of D but not the interior of D. One defines the width of the body as before. If the width of D is the same in all directions, then one says that the body is of constant width and calls its boundary a surface of constant width, and the body itself is referred to as a *spheroform*.

A sphere is obviously a surface of constant width. Contrary to common belief the Reuleaux tetrahedron is *not* a surface of constant width. However, there are two different ways of smoothing subsets of the edges of the Reuleaux tetrahedron to form Meissner tetrahedra, surfaces of constant width which were conjectured by Bonnesesn & Fenchel (1934) to have the minimum volume among all shapes with the same constant width; this conjecture remains unsolved. Among all surfaces of revolution with the same constant width, the one with minimum volume is the shape swept out by a Reuleaux triangle rotating about one of its axes of symmetry (Campi, Colesanti & Gronchi 1996).

Projective Geometry

In mathematics, projective geometry is the study of geometric properties that are invariant under projective transformations. This means that, compared to elementary geometry, projective geometry has a different setting, projective space, and a selective set of basic geometric concepts.

The basic intuitions are that projective space has *more* points than Euclidean space, in a given dimension, and that geometric transformations are permitted that move the extra points (called "points at infinity") to traditional points, and vice versa.

Properties meaningful in projective geometry are respected by this new idea of transformation, which is more radical in its effects than expressible by a transformation matrix and translations (the affine transformations).

The first issue for geometres is what kind of geometric language is adequate to the novel situation? It is not possible to talk about angles in projective geometry as it is in Euclidean geometry, because angle is an example of a concept not invariant under projective transformations, as is seen clearly in perspective drawing. One source for projective geometry was indeed the theory of perspective. Another difference from elementary geometry is the way in which parallel lines can be said to meet in a point at infinity, once the concept is translated into projective geometry's terms. Again this notion has an intuitive basis, such as railway tracks meeting at the horizon in a perspective drawing.

While the ideas were available earlier, projective geometry was mainly a development of the nineteenth century. A huge body of research made it the most representative field of geometry of that time. This was the theory of complex projective space, since the coordinates used (homogeneous coordinates) were complex numbers. Several major strands of more abstract mathematics (including invariant theory, the Italian school of algebraic geometry, and Felix Klein's Erlangen programme leading to the study of the classical groups) built on projective geometry. It was also a subject with a large number of practitioners for its own sake, under the banner of synthetic geometry. Another field that emerged from axiomatic studies of projective geometry is finite geometry.

The field of projective geometry is itself now divided into many research subfields, two examples of which are projective algebraic geometry (the study of projective varieties) and projective differential geometry (the study of differential invariants of the projective transformations).

Overview

Projective geometry is an elementary non-metrical form of geometry, meaning that it is not based on a concept of distance. In two dimensions it begins with the study of configurations of points and lines. That there is indeed some geometric interest in this sparse setting was seen as projective geometry was developed by Desargues and others in their exploration of the principles of perspective art. In higher dimensional spaces there are considered hyperplanes (that always meet), and other linear subspaces, which exhibit the principle of duality. The simplest illustration of duality is in the projective plane, where the statements "two distinct points determine a unique line" (i.e. the line through them) and "two distinct lines determine a unique point" (i.e. their point of intersection) show the same structure as propositions. Projective geometry can also be seen as a geometry of constructions with a straight-edge alone. Since projective geometry excludes compass constructions, there are no circles, no angles, no measurements, no parallels, and no concept of intermediacy. It was realised that the theorems that do hold in projective geometry are simpler statements. For example the different conic sections are all equivalent in (complex) projective geometry, and some theorems about circles can be seen as special cases of these general theorems.

In the early 19th century the work of Poncelet, Lazare Carnot and others established projective geometry as an independent field of mathematics . Its rigorous foundations were addressed by Karl von Staudt and perfected by Italians Giuseppe Peano, Mario Pieri, Alessandro Padoa and Gino Fano late in the 19th century. Projective geometry, like affine and Euclidean geometry, can also be developed from the Erlangen programme of Felix Klein; projective geometry is characterised by invariants under transformations of the projective group.

After much work on the very large number of theorems in the subject, therefore, the basics of projective geometry became understood. The incidence structure and the cross-ratio are fundamental invariants under projective transformations. Projective geometry can be modeled by the affine plane (or affine space) plus a line (hyperplane) "at infinity" and then treating that line (or hyperplane) as "ordinary". An algebraic model for doing projective geometry in the style of analytic geometry is given by homogeneous coordinates. On the other hand axiomatic studies revealed the existence of non-Desarguesian planes, examples to show that the axioms of incidence can be modelled (in two dimensions only) by structures not accessible to reasoning through homogeneous coordinate systems.

In a foundational sense, projective geometry and ordered geometry are elementary since they involve a minimum of axioms and either can be used as the foundation for affine and Euclidean geometry. Projective geometry is not "ordered" and so it is a distinct foundation for geometry.

History

The first geometrical properties of a projective nature were discovered in the third century by Pappus of Alexandria. Filippo Brunelleschi (1404–1472) started investigating the geometry of perspective in 1425. Johannes Kepler (1571–1630) and Gérard Desargues (1591–1661) independently developed the pivotal concept of the "point at infinity". Desargues developed an alternative way of constructing perspective drawings by generalising the use of vanishing points to include the case when these are infinitely far away. He made Euclidean geometry, where parallel lines are truly parallel, into a special case of an all-encompassing geometric system. Desargues's study on conic sections drew the attention of 16-year old Blaise Pascal and helped him formulate Pascal's theorem. The works of Gaspard Monge at the end of 18th and beginning of 19th century were important for the subsequent development of projective geometry. The work of Desargues was ignored until Michel Chasles chanced upon a handwritten copy in 1845. Meanwhile, Jean-Victor Poncelet had published the foundational treatise on projective geometry in 1822. Poncelet separated the projective properties of objects in individual class and establishing a relationship between metric and projective properties. The non-Euclidean geometries discovered shortly thereafter were eventually demonstrated to have models, such as the Klein model of hyperbolic space, relating to projective geometry.

This early 19th century projective geometry was a stepping stone from analytic geometry to algebraic geometry. When treated in terms of homogeneous coordinates, projective geometry looks like an extension or technical improvement of the use of coordinates to reduce geometric problems to algebra, an extension reducing the number of special cases. The detailed study of quadrics and the "line geometry" of Julius Plücker still form a rich set of examples for geometres working with more general concepts.

The work of Poncelet, Steiner and others was not intended to extend analytic geometry. Techniques were supposed to be *synthetic*: in effect projective space as now understood was to be introduced axiomatically. As a result, reformulating early work in projective geometry so that it satisfies current standards of rigor can be somewhat difficult. Even in the case of the projective plane alone, the axiomatic approach can result in models not describable via linear algebra.

This period in geometry was overtaken by research on the general algebraic curve by Clebsch, Riemann, Max Noether and others, which stretched existing techniques, and then by invariant theory. Towards the end of the century the Italian school of algebraic geometry (Enriques, Segre, Severi) broke out of the traditional subject matter into an area demanding deeper techniques.

In the later part of the 19th century, the detailed study of projective geometry became less fashionable, although the literature is voluminous. Some important work was done in enumerative geometry in particular, by Schubert, that is now seen as anticipating the theory of Chern classes, taken as representing the algebraic topology of Grassmannians.

Paul Dirac studied projective geometry and used it as a basis for developing his concepts of Quantum Mechanics, although his published results were always in algebraic form. See a blog article referring to an article and a book on this subject, also to a talk Dirac gave to a general audience in 1972 in Boston about projective geometry, without specifics as to its application in his physics.

Description

Projective geometry is less restrictive than either Euclidean geometry or affine geometry. It is an intrinsically non-metrical geometry, whose facts are independent of any metric structure. Under the projective transformations, the incidence structure and the cross-ratio are preserved. Projective geometry formalises one of the central principles of perspective art: that parallel lines meet at infinity, and therefore are drawn that way. In essence, a projective geometry may be thought of as an extension of Euclidean geometry in which the "direction" of each line is subsumed within the line as an extra "point", and in which a "horizon" of directions corresponding to coplanar lines is regarded as a "line". Thus, two parallel lines meet on a horizon line in virtue of their possessing the same direction.

Idealized directions are referred to as points at infinity, while idealised horizons are referred to as lines at infinity. In turn, all these lines lie in the plane at infinity. However, infinity is a metric concept, so a purely projective geometry does not single out any points, lines or plane in this regard—those at infinity are treated just like any others.

Because a Euclidean geometry is contained within a Projective geometry, with Projective geometry having a simpler foundation, general results in Euclidean geometry may be arrived at in a more transparent fashion, where separate but similar theorems in Euclidean geometry may be handled collectively within the framework of projective geometry. For example, parallel and nonparallel lines need not be treated as separate cases – we single out some arbitrary projective plane as the ideal plane and locate it "at infinity" using homogeneous coordinates.

Additional properties of fundamental importance include Desargues' Theorem and the Theorem of Pappus. In projective spaces of dimension 3 or greater there is a construction that allows one to prove Desargues' Theorem. But for dimension 2, it must be separately postulated.

Under Desargues' Theorem, combined with the other axioms, it is possible to define the basic operations of arithmetic, geometrically. The resulting operations satisfy the axioms of a field—except that the commutativity of multiplication requires Pappus's hexagon theorem. As a result, the points of each line are in one to one correspondence with a given field, F, supplemented by an additional element, W, such that rW = W, "W = W, r+W = W, r/0 = W, r/W = 0, W"r = r"W = W. However, 0/0, W/W, W+W, W"W, 0W and W0 remain undefined.

Projective geometry also includes a full theory of conic sections, a subject already very well developed in Euclidean geometry. There are clear advantages in being able to think of a hyperbola and an ellipse as distinguished only by the way the hyperbola *lies across the line at infinity*; and that a parabola is distinguished only by being tangent to the same line.

The whole family of circles can be seen as *conics passing through two given points on the line at infinity*—at the cost of requiring complex coordinates.

Since coordinates are not "synthetic", one replaces them by fixing a line and two points on it, and considering the *linear system* of all conics passing through those points as the basic object of study. This approach proved very attractive to talented geometres, and the field was thoroughly worked over. An example of this approach is the multi-volume treatise by H. F. Baker.

There are many projective geometries, which may be divided into discrete and continuous: a *discrete* geometry comprises a set of points, which may or may not be *finite* in number, while a *continuous* geometry has infinitely many points with no gaps in between.

The only projective geometry of dimension 0 is a single point. A projective geometry of dimension 1 consists of a single line containing at least 3 points. The geometric construction of arithmetic operations cannot be carried out in either of these cases. For dimension 2, there is a rich structure in virtue of the absence of Desargues' Theorem.

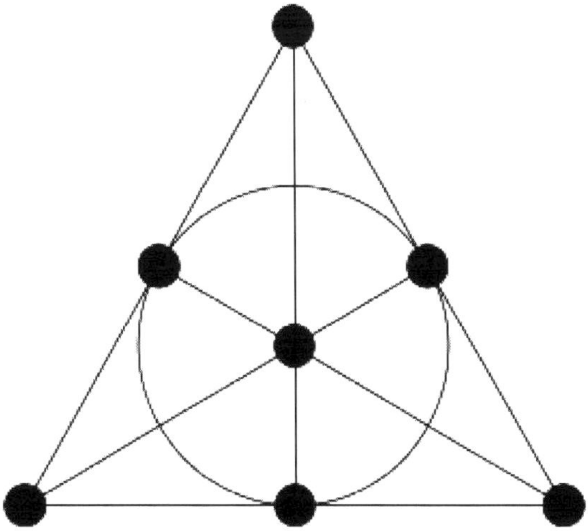

Figure: *The Fano plane is the projective plane with the fewest points and lines.*

According to Greenberg (1999) and others, the simplest 2-dimensional projective geometry is the Fano plane, which has 3 points on every line, with 7 points and lines in all arranged with the following schedule of collinearities:

- [ABC]
- [ADE]
- [AFG]
- [BDG]
- [BEF]
- [CDF]
- [CEG]

with the affine coordinates A = {0,0}, B = {0,1}, C = {0,W} = {1,W}, D = {1,0}, E = {W,0} = {W,1}, F = {1,1}, G = {W, W}. The coordinates in a Desarguesian plane for the points designated to be the points at infinity (in this example: C, E and G) are generally not unambiguously defined.

In standard notation, a finite projective geometry is written PG(a,b) where:

a is the projective (or geometric) dimension, and

b is one less than the number of points on a line (called the *order* of the geometry).

Thus, the example having only 7 points is written PG(2,2).

The term "projective geometry" is sometimes used to indicate the generalised underlying abstract geometry, and sometimes to indicate a particular geometry of wide interest, such as the metric geometry of flat space which we analyse through the use of homogeneous coordinates, and in which Euclidean geometry may be embedded (hence its name, Extended Euclidean plane).

The fundamental property that singles out all projective geometries is the *elliptic* incidence property that any two distinct lines L and M in the projective plane intersect at exactly one point P.

The special case in analytic geometry of *parallel* lines is subsumed in the smoother form of a line *at infinity* on which P lies. The *line at infinity* is thus a line like any other in the theory: it is in no way special or distinguished. (In the later spirit of the Erlangen programme one could point to the way the group of transformations can move any line to the *line at infinity*).

Given a line l and a point P not on the line, the elliptic parallel property contrasts with the Euclidean and hyperbolic parallel properties as follows:

Elliptic: any line through P meets l in just one point.

Euclidean: just one line through P may be found, which does not meet l.

Hyperbolic: more than one line through P may be found, which do not meet l.

The elliptic parallel property is the key idea which leads to the principle of projective duality, possibly the most important property which all projective geometries have in common.

Duality

In 1825, Joseph Gergonne noted the principle of duality characterising projective plane geometry: given any theorem or definition of that geometry, substituting *point* for *line*, *lie on* for *pass through*, *collinear* for *concurrent*, *intersection* for *join*, or vice versa, results in another theorem or valid definition, the "dual" of the first. Similarly in 3 dimensions, the duality relation holds between points and planes, allowing any theorem to be transformed by swapping *point* and *plane, is contained by* and *contains.* More generally, for projective spaces of dimension N, there is a duality between the subspaces of dimension R and dimension N–R–1. For N = 2, this specialises to the most commonly known form of duality—that between points and lines. The duality principle was also discovered independently by Jean-Victor Poncelet.

To establish duality only requires establishing theorems which are the dual versions of the axioms for the dimension in question. Thus, for 3-dimensional spaces, one needs to show that (1*) every point lies in 3 distinct planes, (2*) every two planes intersect in a unique line and a dual version of (3*) to the effect: if the intersection of plane P and Q is coplanar with the intersection of plane R and S, then so are the respective intersections of planes P and R, Q and S (assuming planes P and S are distinct from Q and R).

In practice, the principle of duality allows us to set up a *dual correspondence* between two geometric constructions. The most famous of these is the polarity or reciprocity of two figures in a conic curve (in 2 dimensions) or a quadric surface (in 3 dimensions). A commonplace example is found in the reciprocation of a symmetrical polyhedron in a concentric sphere to obtain the dual polyhedron.

Axioms of Projective Geometry

Any given geometry may be deduced from an appropriate set of axioms. Projective geometries are characterised by the "elliptic parallel" axiom, that *any two planes always meet in just one line*, or in the plane, *any two lines always meet in just one point.* In other words, there are no such things as parallel lines or planes in projective geometry. Many alternative sets of axioms for projective geometry have been proposed.

Whitehead's Axioms

These axioms are based on Whitehead, "The Axioms of Projective Geometry". There are two types, points and lines, and one "incidence" relation between points and lines. The three axioms are:

- *G1:* Every line contains at least 3 points
- *G2:* Every two points, A and B, lie on a unique line, AB.
- *G3:* If lines AB and CD intersect, then so do lines AC and BD (where it is assumed that A and D are distinct from B and C).

The reason each line is assumed to contain at least 3 points is to eliminate some degenerate cases. The spaces satisfying these three axioms either have at most one line, or are projective spaces of some dimension over a division ring, or are non-Desarguesian planes.

One can add further axioms restricting the dimension or the coordinate ring. For example, Coxeter's *Projective Geometry*, references Veblen in the three axioms above, together with a further 5 axioms that make the dimension 3 and the coordinate ring a commutative field of characteristic not 2.

Axioms Using a Ternary Relation

One can pursue axiomatisation by postulating a ternary relation, [ABC] to denote when three points (not all necessarily distinct) are collinear. An axiomatisation may be written down in terms of this relation as well:

- C0: [ABA]
- C1: If A and B are two points such that [ABC] and [ABD] then [BDC]
- C2: If A and B are two points then there is a third point C such that [ABC]
- C3: If A and C are two points, B and D also, with [BCE], [ADE] but not [ABE] then there is a point F such that [ACF] and [BDF].

For two different points, A and B, the line AB is defined as consisting of all points C for which [ABC]. The axioms C0 and C1 then provide a formalisation of G2; C2 for G1 and C3 for G3.

The concept of line generalises to planes and higher dimensional subspaces. A subspace, AB...XY may thus be recursively defined in terms of the subspace AB...X as that containing all the points of all lines YZ, as Z ranges over AB...X. Collinearity then generalises to the relation of "independence". A set {A, B, ..., Z} of points is independent, [AB...Z] if {A, B, ..., Z} is a minimal generating subset for the subspace AB...Z.

The projective axioms may be supplemented by further axioms postulating limits on the dimension of the space. The minimum dimension is determined by the existence of an independent set of the required size. For the lowest dimensions, the relevant conditions may be stated in equivalent form as follows. A projective space is of:

- (L1) at least dimension 0 if it has at least 1 point,
- (L2) at least dimension 1 if it has at least 2 distinct points (and therefore a line),
- (L3) at least dimension 2 if it has at least 3 non-collinear points (or two lines, or a line and a point not on the line),
- (L4) at least dimension 3 if it has at least 4 non-coplanar points.

The maximum dimension may also be determined in a similar fashion. For the lowest dimensions, they take on the following forms. A projective space is of:

- (M1) at most dimension 0 if it has no more than 1 point,
- (M2) at most dimension 1 if it has no more than 1 line,
- (M3) at most dimension 2 if it has no more than 1 plane,

and so on. It is a general theorem (a consequence of axiom (3)) that all coplanar lines intersect—the very principle Projective Geometry was originally intended to embody. Therefore, property (M3) may be equivalently stated that all lines intersect one another.

It is generally assumed that projective spaces are of at least dimension 2.

In some cases, if the focus is on projective planes, a variant of M3 may be postulated. The axioms of (Eves 1997: 111), for instance, include (1), (2), (L3) and (M3). Axiom (3) becomes vacuously true under (M3) and is therefore not needed in this context.

Axioms for Projective Planes

In incidence geometry, most authors give a treatment that embraces the Fano plane PG(2, 2) as the minimal finite projective plane. An axiom system that achieves this is as follows:

- (P1) Any two distinct points lie on a unique line.
- (P2) Any two distinct lines meet in a unique point.
- (P3) There exist at least four points of which no three are collinear.

Coxeter's *Introduction to Geometry* gives a list of five axioms for a more restrictive concept of a projective plane attributed to Bachmann, adding Pappus's theorem to the list of axioms above (which eliminates non-Desarguesian planes) and excluding projective planes over fields of characteristic 2 (those that don't satisfy Fano's axiom). The restricted planes given in this manner more closely resemble the real projective plane.

Hilbert's Third Problem

The third on Hilbert's list of mathematical problems, presented in 1900, is the easiest one. The problem is related to the following question: given any two polyhedra of equal volume, is it always possible to cut the first into finitely many polyhedral pieces which can be reassembled to yield the second? Based on earlier writings by Gauss, Hilbert conjectured that this is not always possible. This was confirmed within the year by his student Max Dehn, who proved that the answer in general is "no" by producing a counterexample.

The answer for the analogous question about polygons in 2 dimensions is "yes" and had been known for a long time; this is the Bolyai–Gerwien theorem.

Dandelin Spheres

In geometry, the Dandelin spheres are one or two spheres that are tangent both to a plane and to a cone that intersects the plane. The intersection of the cone and the plane is a conic section, and the point at which either sphere touches the plane is a focus of the conic section, so the Dandelin spheres are also sometimes called focal spheres.

The Dandelin spheres were discovered in 1822. They are named in honour of the Belgian mathematician Germinal Pierre Dandelin, though Adolphe Quetelet is sometimes given partial credit as well. The Dandelin spheres can be used to prove at least two important theorems. Both of those theorems were known for centuries before Dandelin, but he made it easier to prove them. The first theorem is that a closed conic section (i.e. an ellipse) is the locus of points such that the sum of the distances to two fixed points (the foci) is constant. This was known to Ancient Greek mathematicians such as Apollonius of Perga, but the Dandelin spheres facilitate the proof.

The second theorem is that for any conic section, the distance from a fixed point (the focus) is proportional to the distance from a fixed line (the directrix), the constant of proportionality being called the eccentricity. Again, this theorem was known to the Ancient Greeks, such as Pappus of Alexandria, but the Dandelin spheres facilitate the proof.

A conic section has one Dandelin sphere for each focus. In particular, an ellipse has two Dandelin spheres, both touching the same nappe of the cone. A hyperbola has two Dandelin spheres, touching opposite nappes of the cone. A parabola has just one Dandelin sphere.

Proof That the Curve has Constant Sum of Distances to Foci

Consider the illustration, depicting a plane intersecting a cone to form an ellipse (the interior of the ellipse is coloured light blue). The two Dandelin spheres are shown, one (G1) above the ellipse, and one (G2) below. The intersection of each sphere with the cone is a circle (coloured white).

- Each sphere touches the plane at a point, and let us call those two points F_1 and F_2.
- Let P be a typical point on the ellipse.
- *Prove:* The sum of distances $d(F_1, P) + d(F_2, P)$ remain constant as the point P moves along the curve.
    - o A line passing through P and the vertex S of the cone intersects the two circles at points P_1 and P_2.
    - o As P moves along the ellipse, P_1 and P_2 move along the two circles.
    - o The distance from F_1 to P is the same as the distance from P_1 to P, *because lines PF_1 and PP_1 are both tangent to the same sphere (G1).*
    - o Likewise, the distance from F_2 to P is the same as the distance from P_2 to P, *because lines PF_2 and PP_2 are both tangent to the same sphere (G2).*

- o Consequently, the sum of distances $d(F_1, P) + d(F_2, P)$ must be constant as P moves along the curve because the sum of distances $d(P_1, P) + d(P_2, P)$ also remains constant.
  - § This follows from the fact that P lies on the straight line from P_1 to P_2, and the distance from P_1 to P_2 remains constant.

This proves a result that had been proved in a different manner by Apollonius of Perga.

If (as is often done) one takes the definition of the ellipse to be the locus of points P such that $d(F_1, P) + d(F_2, P)$ = a constant, then the argument above proves that the intersection of a plane with a cone is indeed an ellipse. That the intersection of the plane with the cone is symmetric about the perpendicular bisector of the line through F_1 and F_2 may be counterintuitive, but this argument makes it clear.

Adaptations of this argument work for hyperbolas and parabolas as intersections of a plane with a cone. Another adaptation works for an ellipse realised as the intersection of a plane with a right circular cylinder.

Proof of the Focus-directrix Property

The directrix of a conic section can be found using Dandelin's construction. Each Dandelin sphere intersects the cone at a circle; let both of these circles define their own planes. The intersections of these two planes with the conic section's plane will be two parallel lines; these lines are the directrices of the conic section. However, a parabola has only one Dandelin sphere, and thus has only one directrix.

Using the Dandelin spheres, it can be proved that any conic section is the locus of points for which the distance from a point (focus) is proportional to the distance from the directrix. Ancient Greek mathematicians such as Pappus of Alexandria were aware of this property, but the Dandelin spheres facilitate the proof.

Neither Dandelin nor Quetelet used the Dandelin spheres to prove the focus-directrix property. The first to do so was apparently Pierce Morton in 1829. The focus-directrix property is essential to proving that astronomical objects move along conic sections around the Sun.

Bibliography

Albert, A.: *Regression and the Moore-Penrose Pseudoinverse*, Academic Press, New York, USA, 1972.

Bishop, R. L. and R. J.: *Crittenden Geometry of Manifolds,* Academic Press, New York, USA, 1964.

Bonola, R.: *History of Non-Euclidean Geometry.* Hafner Publishing Company, New York, USA, 1912.

Boothby, W. M.: *An Introduction to Differentiable Manifolds and Riemannian Geometry,* Academic Press, New York, USA, 1975.

Cartan, E.: *Geometry of Riemannian Spaces,* Math Sci Press, Brookline, MA, USA, 1983.

Darling, R. W. R.: *Differential Forms and Connections,* Cambridge University Press, Cambridge, UK, 1994.

Do Carmo, M. P.: *Differential Geometry of Curves and Surfaces,* Prentice Hall, Englewood Cliffs, NJ, USA, 1976.

Doolin, B. F. and C. F. Martin: *Introduction to Differential Geometry for Engineers,* Marcel Dekker, Inc., New York, NY, USA, 1990.

Eisenhart, L. P.: *An Introduction to Differential Geometry with Use of the Tensor Calculus,* Princeton University Press, Princeton, USA, 1940.

Farebrother, R. W.: *Linear Least Squares Computations*, Marcel Dekker Inc., New York, USA, 1988.

Field, Judith V.: *The Geometrical Work of Girard Desargues.* Springer-Verlag, New York, 1987.

Gerretsen, J. C. H.: *Lectures on Tensor Calculus and Differential Geometry,*Academic Press, New York, USA, 1982.

Gray, A.: *Modern Differential Geometry of Curves and Surfaces,* CRC Press, Boca Raton, FL, USA, 1993.

Gruber, M. H. J.: *Regression Estimators: A Comparitive Study*, Academic Press, San Diego CA, 1990.

Kendall, M. G.: *A Course in the Geometry of n Dimensions*, Hafner Publishing Company, New York, USA, 1961.

Levi-Civita, T.: *The Absolute Differential Calculus,* Dover Publications, New York, USA, 1977.

Millman, R. S. and G. D. Parker: *Elements of Differential Geometry,* Dover Publications, New York, USA, 2009.

Mlodinow, M.: *Euclid's window (the story of geometry from parallel lines to hyperspace)*, Harper & Row, New York, USA, 1974.

Murray, M. K. and J. W. Rice: *Differential Geometry and Statistics*, Chapman & Hall, London, UK. 1993.

Nomizu, K. and T. Sasaki: *Affine Differential Geometry,* Cambridge University Press, Cambridge, UK, 1966.

Postnikov, M. M.: *The Variational Theory of Geodesics,* Dover, Publications, New York, 1983.

Robinson, E. A.: *Least Squares Regression Analysis in Terms of Linear Algebra*, Goose Pond Press, Houston, USA, 1981.

Saville, D. J. and G. R. Wood: *Statistical Methods: The Geometric Approach*, Springer-Verlag, New York, USA, 1991.

Weintraub, S. H.: *Differential Forms: A Complement to Vector Calculus,* Academic Press, San Diego, CA, USA, 1997.

Index

A

Abscissa, 5, 8, 9, 85, 86, 95.
Algebraic Definition, 57.
Alternate Viewpoints, 16.
Ambiguities, 179, 198.
Archimedean Spiral, 30, 32, 33, 35.
Astronomical Coordinate, 87, 253.
Axes Conventions, 17, 222.

B

Basis Vectors, 102, 103, 104, 105, 106.
Bijectivity, 59.
Bipolar Coordinates, 87, 88, 89, 90, 91, 92, 111, 112.
Bispherical Coordinates, 88, 91, 92.

C

Cartesian Formulas, 11.
Cartesian Space, 10, 28.
Conic Section, 33, 54, 61, 133, 134, 135, 136, 137, 138, 139, 141, 149, 254, 255, 257, 258, 261, 266, 267, 268.
Coordinate System, 1, 3, 4, 5, 6, 7, 8, 9, 10, 15, 21, 25, 26, 27, 29, 30, 31, 32, 35, 36, 37, 38, 39, 40, 41, 42, 43, 44, 45, 55, 82, 95, 96, 97, 98, 99, 100, 101, 102, 103, 106, 112, 113, 114, 115, 116, 117, 118, 119, 121, 123, 125, 128, 129, 130, 148, 149, 171, 202, 218, 219, 222, 225, 226, 238, 248, 253, 254, 255, 256, 263, 268.

D

Dandelin Spheres, 266, 267, 268.
Dual Theorems, 51.

E

Enriched Circle, 67.
Euclidean Geometry, 3, 46, 68, 124, 142, 144, 145, 146, 148, 149, 150, 151, 152, 153, 154, 155, 157, 158, 160, 235, 243, 250, 256, 258, 260, 261, 262.
Euler Angles, 18, 169, 170, 176, 177, 178, 179, 181, 182, 183, 184, 185, 188, 190, 191, 192, 193, 194, 201, 202, 207, 214, 215, 217, 219, 220, 222, 223.
Euler Characteristic, 68, 81.
Explaining Quaternions, 188.

G

Gaussian Curvature, 68, 115, 238, 239.
Generalizations, 7, 70, 71, 81.
Geographic Coordinates, 44.
Geometric Intuition, 55, 59.
Gimbal Analogy, 183.
Glide Reflection, 11, 12, 13.

H

Homogeneity, 47.
Homogeneous Coordinates, 45, 46, 47, 48, 49, 52, 55, 56, 57, 58, 59, 62, 63, 83, 86, 96, 97, 98, 99, 211, 257, 258, 259, 260, 262.

Hyperbolic Coordinates, 92, 93, 95.
Hyperbolic Geometry, 68, 92, 94, 235.
Hypersphere, 79, 168, 169, 184, 187, 188, 222, 230, 233.

I

Infinite Objects, 152.
Infinitesimal Rotations, 209, 210.
Intermediate Frames, 183.
Isomorphisms, 156, 157.
Isothermal Coordinates, 113, 114.

L

Land Vehicles, 19, 20.
Line Coordinates, 49, 96, 97, 98.
Logical Basis, 152, 155.
Longitudinal Axis, 21, 36, 37, 218, 219.

M

Manifold, 1, 16, 17, 64, 66, 67, 68, 69, 70, 71, 72, 73, 74, 75, 76, 77, 78, 79, 80, 81, 82, 98, 99, 112, 113, 114, 115, 116, 156, 157, 162, 164, 169, 205, 207, 208, 230, 231, 235.
Measuring Angles, 243.
Menaechmus, 134.
Multilinear Algebra, 16.

P

Patchwork, 74, 75.
Physical Science, 93.
Plane Geometry, 10, 143, 144, 235, 263.
Plücker Coordinates, 55, 57, 58, 59, 60, 61, 83, 98.
Polar Coordinate, 4, 29, 30, 31, 34, 35, 36, 37, 38, 41, 45, 83, 87, 88, 89, 90, 91, 92, 99, 100, 101, 111, 112, 116, 183, 222.,
Polar Equation, 31, 32, 33, 35.
Projective Geometry, 46, 49, 51, 56, 62, 135, 141, 149, 158, 160, 255, 256, 257, 258, 259, 260, 261, 262, 264.

R

Rotation Formalisms, 171.
Rotation Matrix , 13, 169, 172, 173, 177, 184, 190, 193, 195, 196, 198, 199, 200, 201, 202, 203, 204, 206, 207, 208, 209, 210, 212, 213, 214, 215, 216, 219.
Rotation Vector, 174, 175, 177, 223, 224.

S

Sea Vehicles, 19, 20.
Slope Definitions, 125.
Solid Geometry, 9, 52, 143.
Space Navigation, 21.
Spatial Rotation, 171, 185, 208.
Spherical Coordinate, 35, 39, 40, 41, 42, 43, 44, 45, 88, 91, 92, 102, 116, 225, 228.
Spherical Geometry, 157, 235, 236.
Stereographic Projection, 157, 230, 232.
Surface Area, 224, 227, 228, 229, 238.

T

Terminology, 21, 68, 147, 229.
Topological Manifolds, 70, 74, 77.
Trigonometric Functions, 121, 122, 123, 125, 127, 128, 129, 130, 132, 204, 242, 245.

U

Uniform Random, 216.
Unique Coordinates, 41.
Unique Cylindrical, 37.

V

Vector Algebra, 105.
Vector Calculus, 107, 45.

□□□